Reporting the delta

Delta Phase 1

III

OK
OK

De lage landen

IX

...En de zee was niet meer

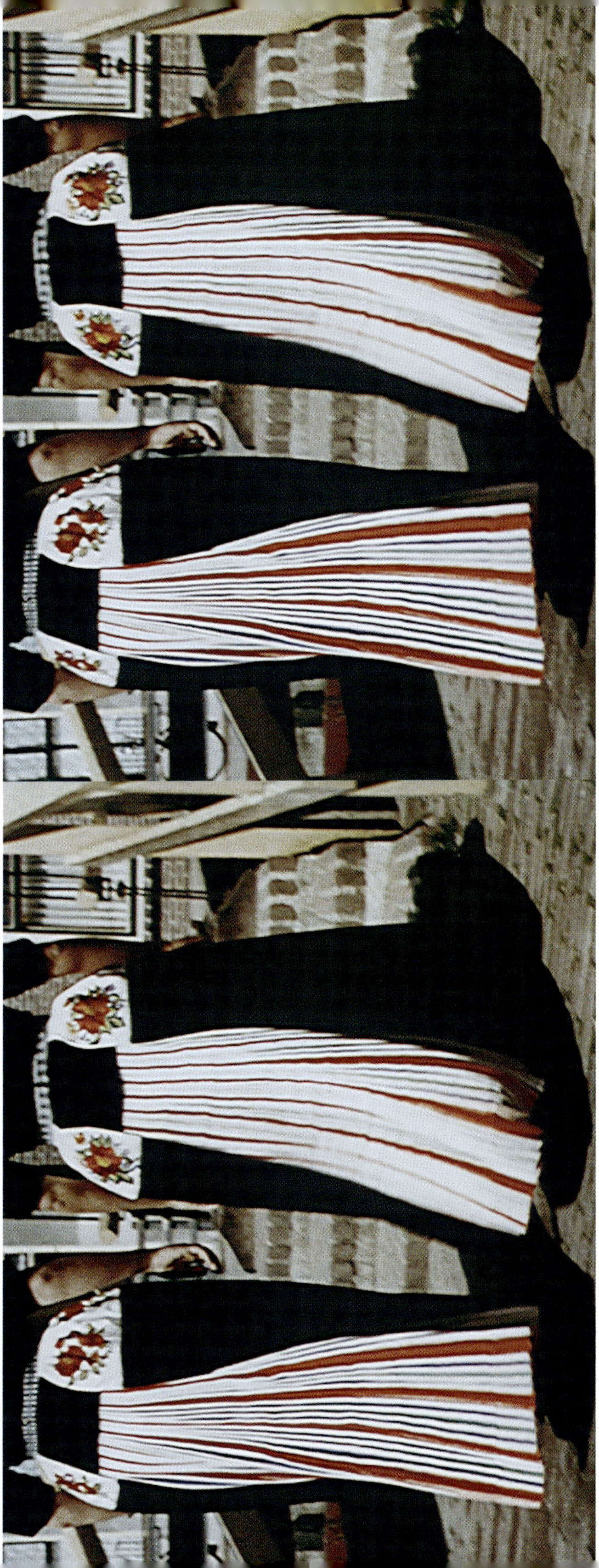

Een nieuw dorp op nieuw land

NAGELE
DORP
IN AANBOUW

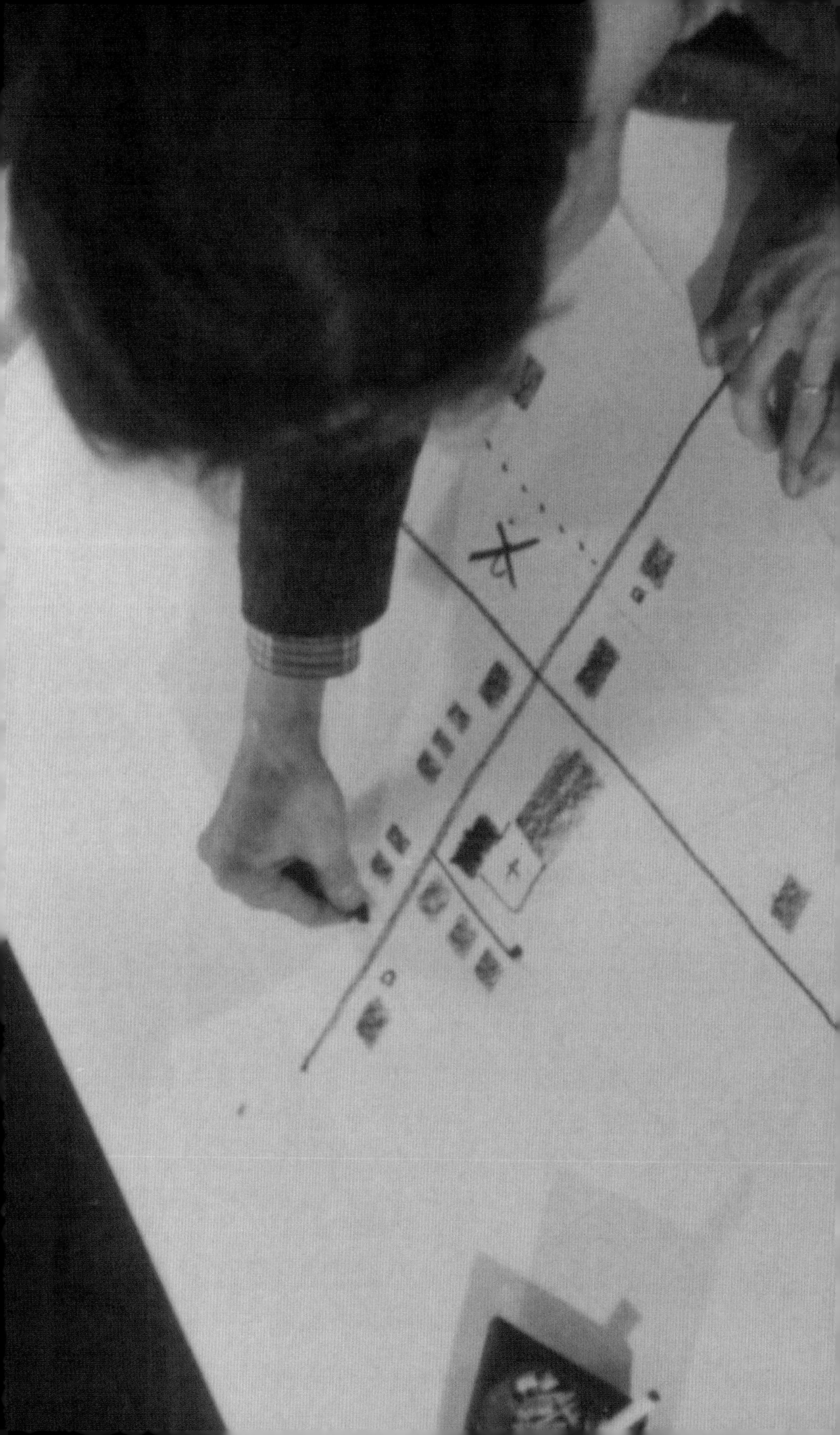

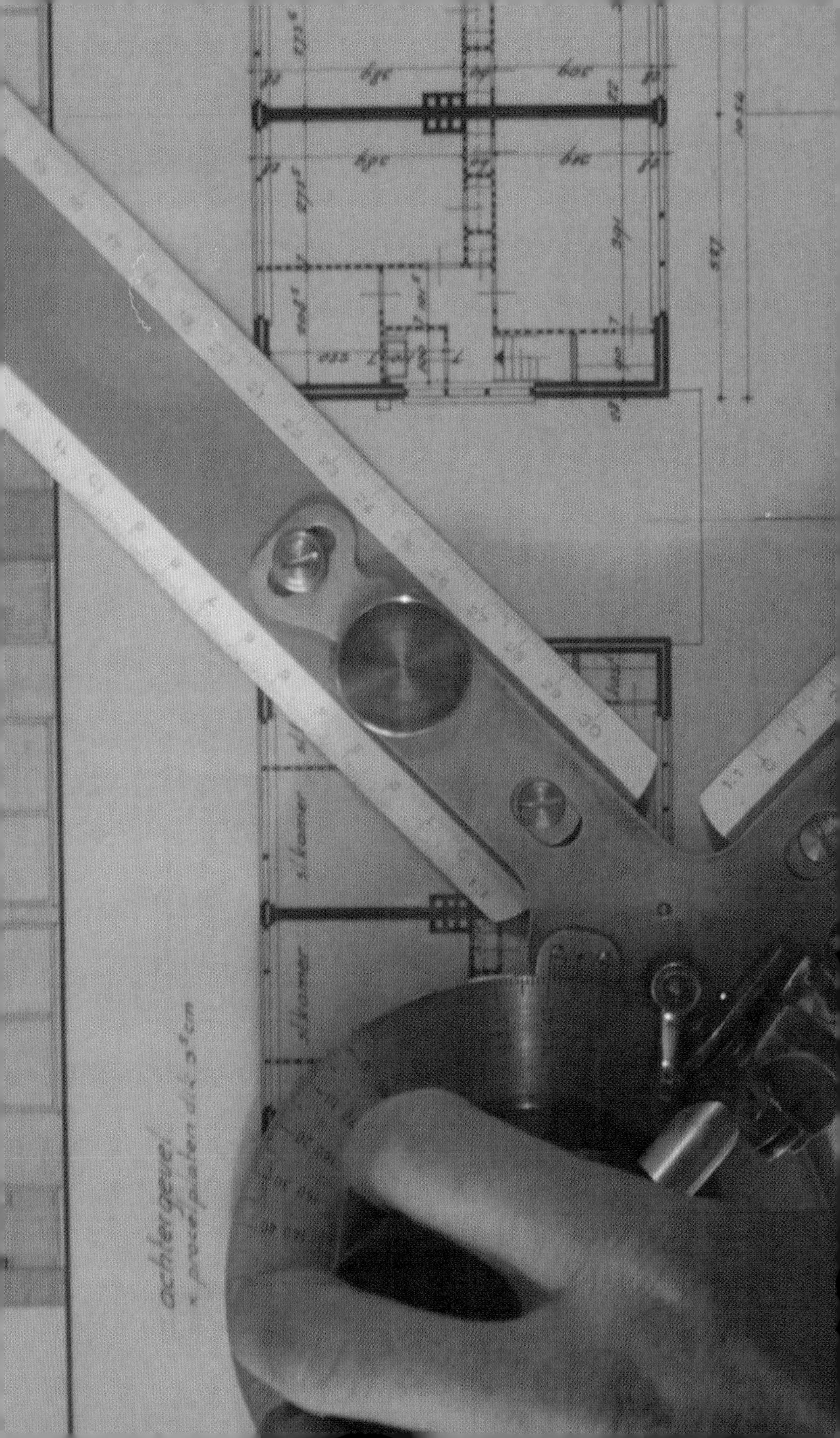
achtergevel

Wieringermeer

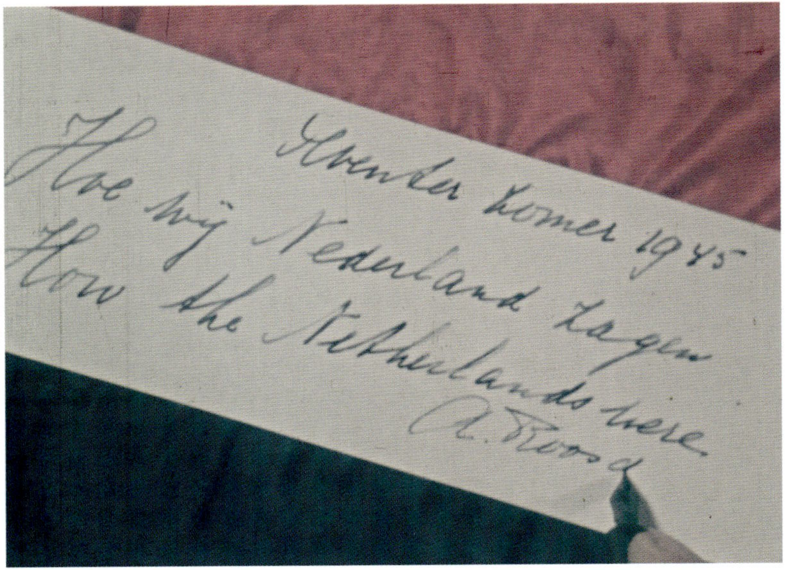
zomer 1945
Hoe wij Nederland zagen
How the Netherlands were

XLIX

JOH.WOERTMAN.

K.N.A.C.
DIJK

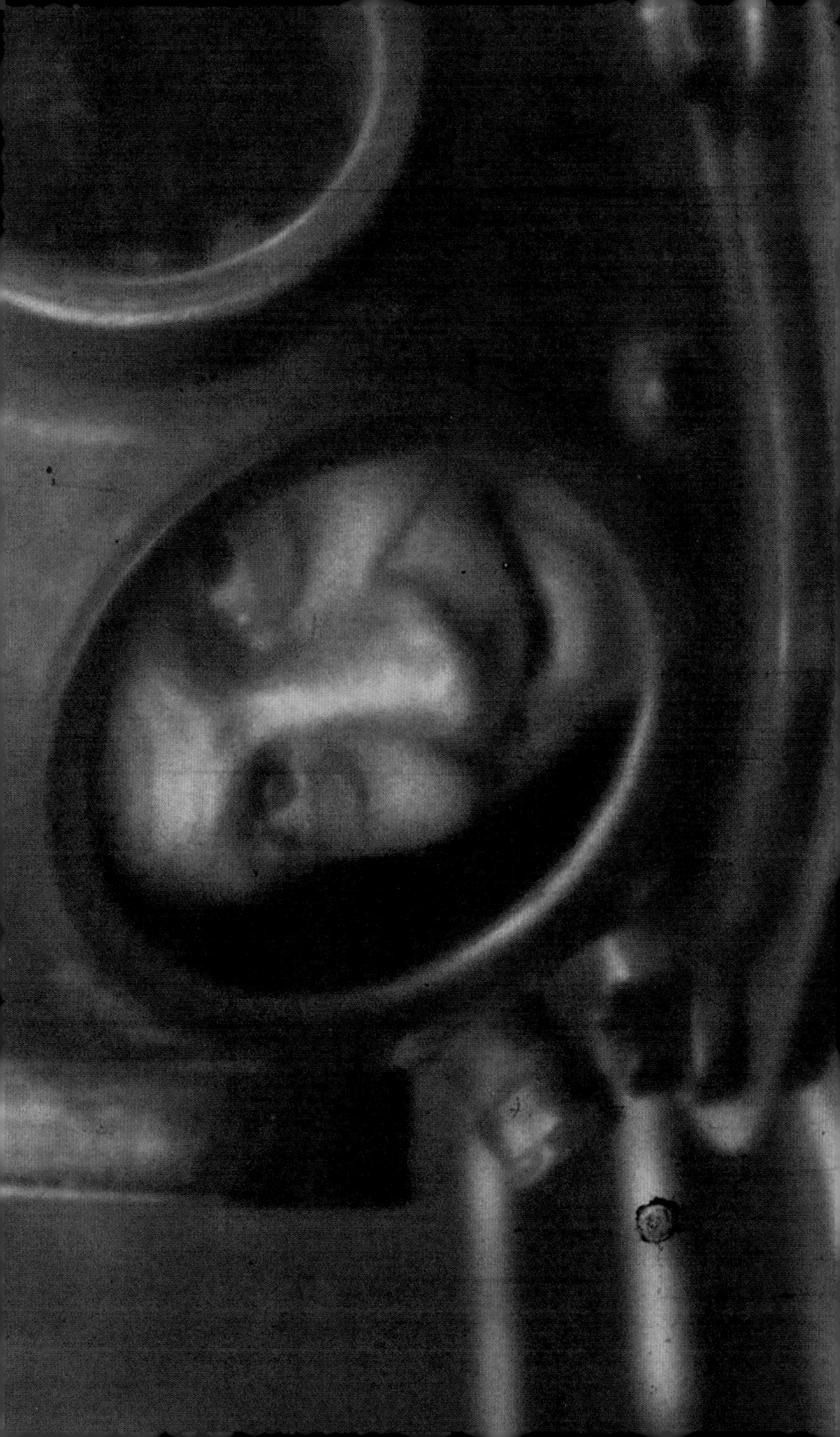

Walcheren

LXI

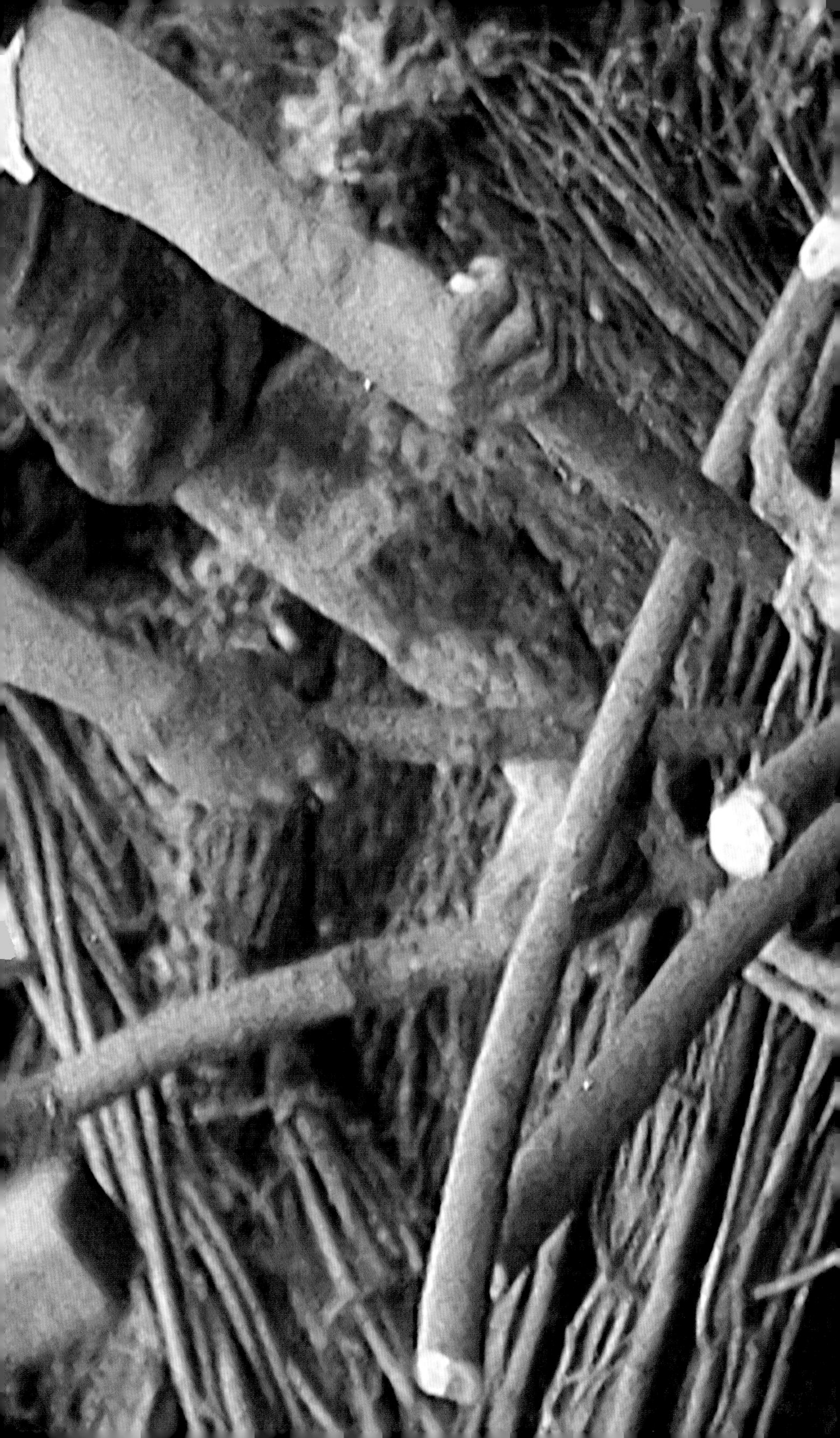

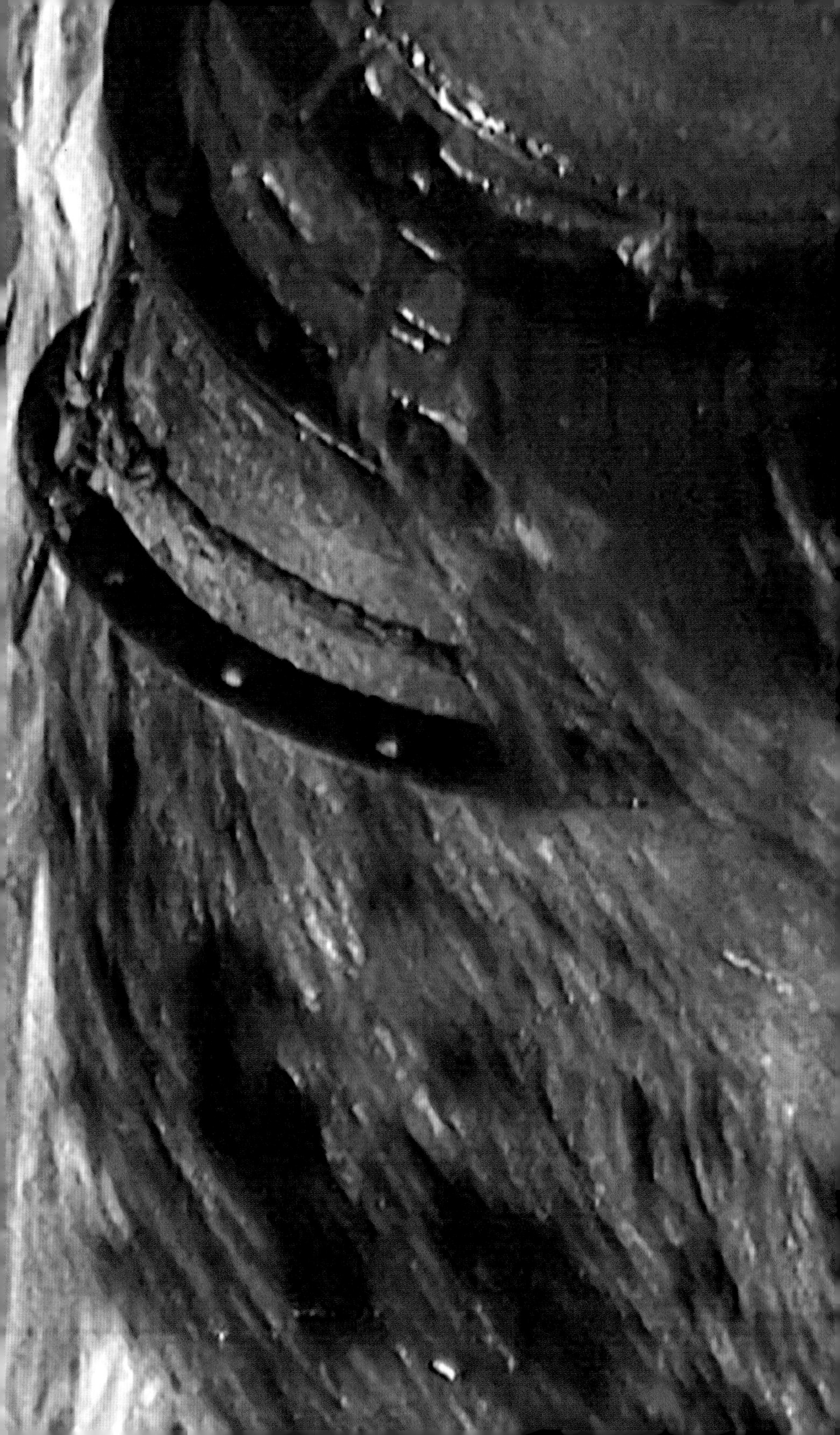

I Stills from *Delta Phase 1*, Bert Haanstra, 1962, Bert Haanstra Films. Eye Filmmuseum. © No further use allowed.

VIII Stills from *De lage landen*, George Sluizer, 1960, Shell Film Unit. Eye Filmmuseum. © No further use allowed.

XX Stills from *…En de zee was niet meer*, Bert Haanstra, 1955, Bert Haanstra Films. Eye Filmmuseum. © No further use allowed.

XXX Stills from *Een nieuw dorp op nieuw land*, Louis van Gasteren, 1960, Spectrum Film. Eye Filmmuseum. © No further use allowed.

XLII Stills from *Wieringermeer*, Alex Roosdorp, 1938, Marofilm. Eye Filmmuseum. © No further use allowed.

XLVIII Stills from *Herwinnen door werken*, Alex Roosdorp and Marie van den Berg, 1945. Marofilm. Eye Filmmuseum. © No further use allowed.

LX Stills from *Walcheren*, Jo de Haas and Mannus Franken, 1945, Nederlandse Werkgemeenschap voor Filmproductie. Eye Filmmuseum. © No further use allowed.

Reporting the Delta

An Exploration of Climate, Space, and Society Through Archival Documentaries

Reporting the Delta

An Exploration of Climate, Space, and Society Through Archival Documentaries

Edited by
Luca Iuorio,
Sophia Arbara,
Carissa Champlin

nai010 publishers

Contents

Between October 2023 and May 2024, we took a series of archival documentary screenings on the road, gathering people to reflect on the ways in which landscapes, infrastructures and lives have been shaped, and continue to be shaped, in the Dutch delta by the intertwining of water, technology, climate and society/everyday life/culture. This itinerant film club formed the basis of the research project 'Reporting space, time and everyday life in the delta', a collaborative experiment that unfolded across film, conversation and public debate. The films shown were neither mere historical records nor illustrative backdrops; they were propositions – vivid articulations of how engineering, policy and everyday life are entangled across space and time in the Dutch delta landscape.

The film club was conceived not as a platform for answers, but as a space for exchange, a way to bring science into dialogue with citizens, designers, engineers, activists, to question inherited myths, and to surface alternative perspectives. The discussions that followed each screening were rich, unpredictable and at times moving. And they are carried on now in the written form of essays, interviews and reflections. This book is not the final outcome of the project but a continuation, a documentation of ongoing inquiry into the realities of climate change, the legacies of infrastructure and the modes of knowing we might need in order to face the uncertain future of the Dutch delta.

The project draws inspiration from Henri Lefebvre's Rhythmanalysis, which posits that understanding space requires more than measurement. It demands attunement to the temporal and bodily rhythms that constitute everyday life. Lefebvre's work reminds us that there is no strict division between subjectivity and objectivity, representation and reality, past and present, art and science. What matters is how we read the world, how we pay attention to patterns, silences, ruptures and how we use those observations to imagine otherwise.

At its core, the project rests on three simple yet far-reaching premises. First: that art, however contested or slippery the term may be, is a practice of knowledge production. This is not a claim but a question of how art can contribute to the scientific study of space, society and environmental change. Second: that archives, especially audiovisual ones, hold enormous potential to inspire thinking about delta futures. They not only preserve the past, but they reveal what has been forgotten, neglected or never fully understood while uncovering the hidden patterns of spatial change through motion pictures. Third: that the changing climate demands more than technical solutions. It calls for public engagement, addressing difficult questions, and a rethinking of our reliance on engineering and infrastructures as unquestioned forms of protection or detachment from climate-related risks.

If this project has sometimes been misread and labeled as 'historical', it is because we have indeed been looking backward. But our approach has been speculative rather than archival in the traditional sense. What interests us is not merely a historical account of what happened, but the potential of the past to reveal new ways of seeing. History is approached not as a closed container of facts but a living reservoir of feelings, observations and unanswered questions. Archives are never neutral but hold a memory, waiting to be made visible, reactivated and pondered.

The use of documentary film, itself a historically Dutch tradition that accompanied other Dutch traditions such as constructing dikes and dams, provided a unique lens for this exploration. Through archival footage, voice-over and constructed scenes, documentaries give access to the atmospheres and effects of particular moments in the past, from the early years of the Delta Works – the marvel of hydraulic engineering developed after the catastrophic 1953 North Sea Flood – to later reflections on environmental change and spatial planning. But more importantly, they allow for ambiguity. The documentary discloses a multiplicity of perspectives: those of the filmmakers, the speakers, the viewers, the places themselves, without flattening them into one truth.

In Reporting the Delta, the documentary form became a space where disciplines met without merging, where science could sit alongside sensitivity and where speculation was not a criticism but a method. There is a kind of latent knowledge embedded in documentaries, not only in their facts, but in their pace, tone and framing. In a moment when the ecological crisis is often reduced to data sets and engineering challenges, we argue for the importance of more unstable and intuitive forms of understanding.

This is not to dismiss the role of engineering. On the contrary, the Netherlands owes much of its survival to it. For centuries, the engineer has been celebrated as a national hero, a figure of rationality, mastery and control over nature. Dikes, dams, storm surge barriers, gates, flumes and pumping stations are not only infrastructures; they are symbols of a national narrative built on an enduring quest for adaptation. Yet today, with sea levels rising, precipitation patterns shifting, large-scale subsidence underway and financial constraints tightening, the symbolic significance of these infrastructures is under pressure. Scenarios that would have once been unthinkable like managed retreat, abandonment, radical adaptation and living with water are now under consideration. Still, the dominant reaction remains one of waiting, waiting for the government, for the water boards, for the technical fix as if the climate one day will become predictable and manageable and a new condition of equilibrium will finally be reached.

This edited collection brings together twenty-five contributions from an eclectic group of authors with diverse perspectives, including film directors, cinephiles, scientists, engineers, artists, activists and designers. Composed of four distinct categories – essays, dialogues, carnets and a digest series – the book invites readers to navigate across multiple ways of seeing and understanding the delta. Essays include peer-reviewed articles that provide insights into the project's core themes. Dialogues feature in-depth conversations with film directors, specialists and practitioners, allowing different disciplines and experiences to encounter each other directly. Carnets offer portraits that blend text and image to highlight ongoing design and engineering practices. The digest series, curated by Rommy Albers (Eye Filmmuseum, Amsterdam), offers a selection of textual and visual reflections on the documentaries that frame the book's main themes. This category describes selected films and incorporates still frames to highlight key ideas visually.

Essays, carnets and dialogues populate the four thematic sections of the project with texts from the digests interwoven throughout, adding an additional layer of interpretation to this project. Following the opening digest on the movie *Delta Phase 1* by Bert Haanstra, 1962, the first thematic section begins with an essay by Guido Borelli that establishes the philosophical and theoretical foundations of the project by drawing from the concept of Rhythmanalysis. This initial section grounds the project in a concrete understanding of rhythms and their manifestation in the daily lives of people. It explores the tensions between natural cyclical rhythms and the societal-imposed linear rhythms, tensions intensified by spatial developments, infrastructural interventions and the longstanding attempts to control nature. Through an exploration of social, ecological and infrastructural rhythms, Pınar Şefkatli's essay offers new perspectives on sustainability and climate change. The section concludes with a *Dialogue with Paul Shepheard* who, through personal narratives and philosophical insights, challenges conventional environmental discourses and invites a rethinking of nature, wilderness and the evolving perception of landscape. Following this section, the digest on the movie *De lage landen* (*Hold Back the Sea*) by George Sluizer, 1960, revisits the reclamation of the Netherlands' western delta, linking ideology, education and engineering through moving images.

From the theoretical groundwork of the first section, the narrative transitions to the second section on design and technology: to the making of mechanical landscapes and the shifting strategies for climate adaptation in the Netherlands. Through a series of essays and carnets, this section addresses the impact of climate change and the urgent need to rethink traditional engineering models. Contributions explore a necessary cultural shift in delta management, as discussed in the carnet by Fransje Hooimeijer; advocacy for new approaches to climate adaptation, presented in the essay by Gaby S. Langendijk, and a call for reimagining design thinking, outlined in the carnet by Joost Adriaanse and Ester van de Wiel. Further contributions, including the essay by João Cortesão and Agnès Patuano, challenge established academic and professional silos, pushing for integration between knowledge on climate resilience and health. Lastly, the making of landscapes is discussed in the *Dialogue with Digna Sinke*, where filmmaking is posited as a lifelong exploration of place, memory and change, an act of observing, preserving and connecting personal and collective histories across landscapes and time. The digests placed within this section, on the movies *... En de zee was niet meer* (*... And the Sea Was No More*) by Bert Haanstra, 1955, *Een nieuw dorp op nieuw land* (*A New Village on New Land*) by Louis van Gasteren, 1960, and *Wieringermeer* by Alex Roosdorp, 1938, chart life in transition: from traditional crafts and village customs to modernist settlements and the rise of agricultural mechanization. These films document the everyday realities of change, portraying how spatial transformations unfold at the human scale.

The third section shifts the focus towards critical approaches to environmental, artistic, political, (de)colonial discourses and local narratives, starting with the *Dialogue with Lia Franken-de Vries*. Marilena Mela emphasizes the necessity of learning from critical environmental and decolonial frameworks by engaging with local knowledge, for example in Ameland, the Netherlands while Jan-Philipp Possmann's essay examines how scientific research and artistic practice can intersect. This section also includes a critical examination of international cases such as the Bengal delta, where dominant narratives of modernization have historically overshadowed local knowledge and histories, as explored in the essay by Kamar Ahmad Simon. The fragility of infrastructural systems and the cycles of destruction and reconstruction are evoked in the digests on the movies *Herwinnen door werken* (*Reclaiming by Works*) by Alex Roosdorp and Marie van den Berg, 1945, and *Walcheren* by Jo de Haas and Mannus Franken, 1945, which depict post-war rebuilding efforts and the resilience embedded in Dutch delta landscapes.

The fourth section returns to the role of cinema and archival work, asking how environmental, spatial and scientific discourses intersect through film. Contributions reflect on the dynamic nature of cinematic spaces, as explored in the essay by Nina Bačun; on filmmaking as a research method within spatial and landscape disciplines, examined in the essays by Peter Veer and Saman Seyff; and on the importance of preserving and reactivating audiovisual archives, as discussed in the carnet by Nicoletta Traversa and Giuseppe Ferrari. Essays in this section show how documentary film, engaged beyond recording the world, can open up new ways of perceiving the relationships between landscapes, infrastructures and social life. The digest on the movie *Dijkbouw* (*Dike Construction*) by Bert Haanstra, 1953, exemplifies the postwar cinematic tradition highlighting human labour and engineering forces in the creation of dike infrastructures.

The book concludes with reflections on cinema as both a representation and an active agent of spatial transformation. It includes an essay by Floris Paalman who analyses how audiovisual media have contributed to shaping spatial development through cultural ecologies and ideologies, and a *Dialogue with Jord den Hollander and Dirk Sijmons* who reflect on how filmmaking has witnessed and documented spatial change across decades. The last digest on the movie *Delta Finale* by Joop Burcksen and Ruud Herblot, 1988, closes the digest series by celebrating the culmination of three decades of Delta Works, a symbolic endpoint and cinematic monument to national ambition and spatial transformation.

By giving voice to varied experiences, ideas, visions, questions and doubts, the book aspires to engage a broader audience in a critical dialogue about the future of the Netherlands. The conversations in this book contribute to the Open Science movement in the Netherlands, led by the Dutch Research Council, adopted by local initiatives like the TU Delft Climate Action Programme, and promoted worldwide by UNESCO to accelerate scientific progress for everyone. But what has been gained by activating the collective memory? Stories shared on growing up in the Dutch delta depict common experiences, such as disaster tourism as part of the everyday mundane in the years following the 1953 flood and a youth hallmarked by the heroic feats of a new generation of engineers. Yet, the pride of these heroics transposed onto another geographical context tells a starkly different story of geopolitics, exploitation and ecological collapse. The Bengal delta lays bare the injustices of modern technological approaches to hold back the water at the cost of natural, ecological and traditional socio-economic flows and rhythms that have sustained regions for millennia while connecting people to their environment. These stories not only teach us of traditions and knowledge long gone or drastically changed, they also bring contextual conditionality to the often narrow realities of the present, broadening our view on the delta in the hope that its future will be built on the stilts of collective action and a range of perspectives.

From a methodological perspective, the contributions to this book showcase the diversity of media and forms of reflection used across disciplines. For instance, researchers take a normative stance in their plea for the integration of climate adaptation and health strategies in cities, evidenced by literature and empirical research findings. Documentary filmmaking extends this methodological repertoire by revealing the spatial dynamics that lay hidden in traditional cartographic methods and by communicating scientific argumentation through creative narration. Such methodological pluralism injects discussions on the delta with a continuous self-reflective and experiential contemplation of space, and indeed of place and belonging. It creates a world filled with imaginaries of unknown territories beyond sensory space and set in speculative time horizons. Rhythmanalysis not only draws connections across rhythms in their various spatiotemporal frequencies; it offers a shared vocabulary for our different ways of studying and understanding social, ecological and infrastructural processes. The careful curation of such diffuse forms of reflection in this book hints at a necessary next step in engaging the arts, science and society in multilogue. Surely, if we share a common goal of more adaptive, liveable spaces, new means of bringing together our different ways of grasping the world are needed. Through the themes of art and science, past, present and future, infrastructure and everyday life, what ultimately ties this work together is climate change – not as a technical variable but as a condition that challenges our ways of knowing, organizing and imagining the world. The delta has long been defined by its relationship to water. With this book we ask whether that relationship can be rethought, not only by engineers and planners, but by all of us.

Luca Iuorio, Sophia Arbara, Carissa Champlin

Figure 1: Posters for the screenings of the itinerant film club Reporting space, time and everyday life in the delta. From October 2023 to May 2024. Luca Iuorio, 2023-2024.

Figure 2: Frame of the project Reporting space, time and everyday life in the delta. Luca Iuorio, 2024.

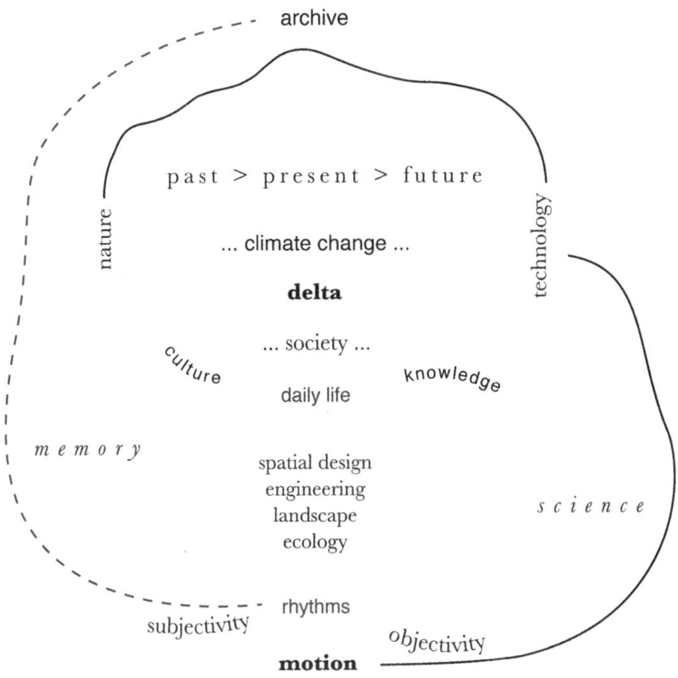

Delta Digests

Rommy Albers

In 1961, Bert Haanstra was commissioned by the Ministry of Transport and Water Management to make the film *Delta Phase 1* about the closure of the Veerse Gat. It would become an iconic film about the Dutch Delta Works – a film that places the project within the heroic struggle of the Dutch against the sea and the encroaching water, and at the same time, reflects on the prosperity the sea has brought us: the sea as the ever-present phenomenon in the life and culture of the Dutch.

The film selection for this book focuses on *Delta Phase 1*. Around it, a number of contemporary films have been chosen that also show the struggle against water and the efforts to control the sea, such as films about the Zuiderzee Works from the first half of the Twentieth Century, about the Delta Works after the flood of 1953 and about the changes these hydraulic engineering projects had on the landscape. They show the centuries-old tradition of reclaiming land from the water, of arranging and rearranging this reclaimed land for agriculture and cattle breeding.

The films show this history and place the Delta Works in a historical context. They legitimize, propagate, inform and show developments and innovations in civil engineering. The films all exude an atmosphere of progress and optimism – even at the expense of age-old traditions. In doing so, they entirely fit in with the atmosphere of reconstruction in the Netherlands after World War II.

All films come from the collections of the Netherlands Institute for Sound and Vision (B&G) and Eye Filmmuseum, two archives that manage and preserve the cinematographic heritage of the Netherlands and ensure that the image of the Netherlands as captured on film remains available for presentation and research.

Delta Phase 1, Bert Haanstra, 1962

Rommy Albers

On 27 April 1961, the Veerse Gat was closed with the placement of the seventh and last caisson. With the completion of the nearly three-kilometre-long dam between Noord-Beveland and Walcheren, the first part of the Delta Works was finished. The Veerse Gat, an open estuary between the two islands, now became a closed inland sea: the Veerse Meer. Veere, the picturesque fishing port, was cut off from the sea and the fishing fleets of Veere and neighbouring Arnemuiden had to find another home port.

The placement of the last caisson was a national event. Queen Juliana was present; dignitaries from both islands symbolically shook hands; and the national and international press covered the spectacle. Of course, film cameras were also present to capture this historic moment. Polygoon-Profilti shot footage for the national newsreel *Hollands Nieuws* and director Bert Haanstra instructed his cameramen to make images for the film *Delta Phase 1*, a public information film commissioned by the Ministry of Transport and Water Management about the closure of the Veerse Gat.

Bert Haanstra was the most famous Dutch national filmmaker at that time. He had won the Grand Prix du Court Métrage at the Cannes Festival with his first short film *Spiegel van Holland* (*Mirror of Holland*) in 1951, an unprecedented success for Dutch film and the start of a successful career. In the 1950s, Haanstra made a number of short documentaries and commissioned films for Shell. His first feature film *Fanfare* (1958) is a benchmark in the history of Dutch feature films. Haanstra's films were highly praised and won him awards at various festivals. The highlight was the Oscar he received in 1960 for his short film *Glas* (*Glass*; 1958). Together with fellow filmmakers such as Herman van der Horst, Ytzen Brusse and Max de Haas, he brought Dutch documentary film to an internationally high level in the 1950s.

It should therefore come as no surprise that the Ministry selected Haanstra for *Delta Phase 1*. He had repeatedly proven his craftsmanship, had the production experience to tackle such a project and, in addition, had already made two films in the past in which he had demonstrated his affinity with the subject. In 1952, he had been commissioned by Shell to make the film *Dijkbouw* (*Dike Construction*; 1953) and three years later Haanstra had directed a film about the fishing villages on the former Zuiderzee: *. . . En de zee was niet meer* (*. . . And the Sea Was No More*; 1955). These two films were about the major hydraulic works carried out in the Netherlands in the first half of the twentieth century: the Zuiderzee Works. These closed off the Zuiderzee, the large inland sea in the centre of the Netherlands, and transformed it into the IJsselmeer, closing it off from the Waddenzee. It was then partly reclaimed to create agricultural land.

Dijkbouw is about a technique developed by Shell to make the construction of dikes faster and cheaper. The footage was shot on a work island in the IJsselmeer, where work was being done on the reclamation of Flevoland. In the film, Haanstra demonstrated that he could show this technical process and provide insight into it in a cinematographic way. To make the film, Haanstra and crew spent a number of weeks on the work island, together with the engineers and workmen.

The other film, *. . . En de zee was niet meer*, is about the other side of the story of the reclamation of the Zuiderzee. It's not about safety, the technical ingenuity of the hydraulic works and the progress they bring, but about the loss and nostalgia experienced by the inhabitants. The closing of the Zuiderzee brought major social changes for the residents of the cities and villages on the Zuiderzee coast. Islands like Urk were connected to the mainland; fishing – the largest source of income for many towns – was lost. The fishing fleets were reduced in size and had to choose other home ports; the prosperous towns became proverbial 'dode steden aan de Zuiderzee' ('dead cities on the Zuiderzee', after the 1922 film by Willy Mullens of the same name). Progress is matched by loss, the optimism of modernity is balanced by a nostalgic view of the past. They are inextricably linked in Haanstra's work.

In January 1961, Bert Haanstra was commissioned by the Ministry to make a film of approximately twenty minutes about the closing of the Veerse Gat between Kamperland (Noord-Beveland) and Breezand on Walcheren. With the construction of this dam, the so-called Drie Eilandenplan (Three Islands Plan) was completed. This plan, developed in the 1930s to connect Walcheren and Noord- and Zuid-Beveland, had become part of the overall Delta Works after the 1953 flood disaster and was completed as the first sub-project of this immense hydraulic engineering project. On 1 October 1960, work on the eastern side of the Veerse Gat, connecting to the Eastern Scheldt, was completed with the commissioning of the Zandkreek Dam. For the closure on the western side – where the Veerse Gat flowed into the North Sea – an important step would be taken in April 1961 with the placement of the caissons that formed the foundation for the Veerse Gat Dam. This sea dam would permanently seal off the Veerse Gat from the North Sea.

The goal was to make a documentary that could be shown as a preliminary film in domestic and foreign cinemas. In addition to a Dutch-spoken version, versions in English, French, German, Spanish, Portuguese and Indonesian were also planned. *Delta Phase 1* was shot in colour on 35 mm, an exception at a time when Dutch film was still almost entirely black and white. For the technical finishing, the British laboratory Technicolor in London was chosen, where Haanstra had already gained experience with the colour films he had made on behalf of Shell.

Delta Phase 1 focuses almost entirely on the placing and sinking of the seven caissons, an event that took place in late April 1961. With multiple cameras on the tugboats and caissons, from the seawalls along the water and from helicopters, Haanstra followed the towing of the caissons from the construction pit near Veere to the new seawall. After the first caisson was placed, the next caissons were placed perpendicular to the previous caisson and pushed into place against the flow, hinged. Once the caissons were in place, they were sunk and stabilized by pouring basalt and stones against the side walls. The open spaces between the caissons were also filled in this way. To avoid variable currents during the operation, the wooden sidings of the caissons were removed, so the water could flow freely through. After the last caisson had been placed on 27 April 1961, steel hatches were lowered using winches, sealing the caissons and permanently closing off the Veerse Gat from the open sea, thus creating the Veerse Meer. The caissons were then filled with stone; sand presses sprayed a total of about four million cubic meters of sand onto the dike body. After a few weeks, the sea dam had acquired the correct profile and was finished with a layer of asphalt bitumen. The road surface was laid and on 23 June 1961, the completion of the first major project of the Delta Works was celebrated with a large-scale theatrical event. Carel Briels, an experienced director of mass plays, staged a spectacle in which 1200 people participated. Haanstra recorded the spectacle from a helicopter and concluded the documentary with this.

Delta Phase 1, just like *Dijkbouw*, gives a detailed picture of the work. However, there is a world of difference in terms of technology. *Dijkbouw* still shows the old, traditional working methods used in the Zuiderzee Works. Cranes pour clay, workers process the clay, install the sheet pile walls and then place large reed mats on which stones are placed to reinforce and strengthen the dam. Everything is on a human scale: materials are delivered in wheelbarrows and laid manually – in stark contrast to the impressive caissons in Zeeland, transported with great precision from construction site to berth with the help of five tugboats.

What both films have in common is the focus on the worker, on those who execute the project. In *Dijkbouw*, they are the anonymous thatchers, stone-layers and crane operators; in *Delta Phase 1* the many workers on the caisson who take care of placement, assembly and closure. Sometimes, however, they also show their faces: the man who meticulously directs the transport of the caissons with the walkie-talkie and the tugboat captain who coordinates the voyage. They are the ones who get the job done, not the higher-ups who shake hands when the last caisson is placed and two islands are permanently connected. To make it more vivid, Haanstra intersected the documentary images with shots taken at the hydraulic engineering laboratory De Voorst in Vollenhoven. There, a scale model of the

Delta Works was made and tested by engineers. They waded through knee-high water to move the miniature caisson, calculating the work and predicting the difficulties.

However, there is an essential distinction between *Delta Phase 1* and *Dijkbouw*. *Dijkbouw* was made entirely in the tradition of the public information film and, above all, shows the work process itself. The work is shown in isolation of its purpose and the surrounding world. This makes the film primarily a kind of 'song of labour', while *Delta Phase 1* is a 'song of the future'. Everything is shown in the context of history and future. The film shows the flood and the Zeeland motto *luctor et emergo* ('I struggle and emerge'): a future is being built, but a future that also brings changes. Zeeland will no longer be as it was. Where previously people dressed in traditional costumes used to watch the fishing fleet sail out, they will now watch the recreational water skiers.

Modernity and Nostalgia

Delta Phase 1 starts with images of 'old, traditional' Zeeland: the church of Biervliet, Zeeland and the women in traditional costumes and traditional customs show the country's ancient culture. However, this culture is also vulnerable, not only because of social changes and advancing technology, but also because of natural hazards and disasters. The recent flood disaster of 1953 and the inundation of Walcheren following an Allied bombing on 3 October 1944 had demonstrated this. The Delta Plan would put an end to the danger posed by the sea. Dams and dikes provided a defence against the advancing sea and storm surges.

But it also meant a social upheaval. Just as when the Zuiderzee was dammed, the natural route to the sea was closed off to large fishing fleets. Haanstra pays extensive attention to this subject and films the last time that Arnemuiden fishermen leave the port of Veere. From then on, they will have their home port in Colijnsplaat on Noord-Beveland. A new fish auction will also be built there. Haanstra shows the grief of the fishermen and residents of Veere and Arnemuiden, but there is understanding that it is inevitable. If Zeeland wants to secure itself, sacrifices will have to be made. Thus, there is also resignation – very different from a quarter of a century earlier during the Zuiderzee Works. Then, there was less understanding and the fishermen reacted vehemently, as shown in films such as *Dood water* (*Dead Water*; 1934) by Gerard Rutten and *De laatste dagen van een eiland* (*The Last Days of an Island*; 1938) by Ernst Winar. In these films, the salinization of fishing waters and the relocation of the fleets lead to mourning and rebellion.

Haanstra identifies loss, but shows it as a natural progression of history. It is part of a process and much good will come in its place. It is this idea of progress that was strongly felt in the Netherlands in the 1950s and 1960s. After World War Two, the Netherlands had enthusiastically begun to repair the material damage caused by the war. Cities were rebuilt, infrastructure repaired. A new, modern country and society were built; our own 'Wirtschafswunder' announced itself. There was a feeling of optimism and confidence in the feasibility of a new Netherlands. Progress took place on the newly reclaimed land in the IJsselmeer: on the island of Wieringen and in the Noordoostpolder. Films such as *Wieringermeer* (Alex Roosdorp; 1938) and *Een nieuw dorp op nieuw land* (*A New Village on New Land*, Louis van Gasteren; 1960) show the new agricultural Netherlands. Small plots of arable and pasture land have been replaced by large, expansive farmland on which the grain is harvested by machines. New residential centres emerge.

The landscape in Zeeland would also change dramatically, not only because of the reallocation of the refertilized land and the mechanization of agriculture, but also because of tourism. In *Delta Phase 1*, we can see that Veere has already been assigned to a new role. The old fishing fleet has disappeared, but a new fleet will return: that of pleasure yachts and sailing boats. Veere will become the centre of the recreational culture that will emerge around the new Veerse Meer. The new bridges and sea walls provide fast connections with the rest of the Netherlands and bungalow parks with holiday homes are built behind the dunes on the islands of Zeeland and Zuid-Holland. In the amateur films of the Mees (1954-55) family we see how in the 1950s people spent their holidays at these new locations.

The Hague and Veere

A year later, on 20 March 1962, *Delta Phase 1* was shown for the first time at a special afternoon screening organized by the Ministry of Transport and Water Management at The Hague's Metropole Theatre. For a large group of invited guests, including workmen who had worked on the project, *Delta Phase 1* was shown together with *Zoo* (1962), another short film by Haanstra. *Zoo* had premiered a week earlier and was shot simultaneously with *Delta Phase 1*. On weekends, when work in Zeeland was at a standstill, Haanstra and his cameraman Fred Tammes and assistant Kees Hin had spied on people and animals in Amsterdam's Artis Zoo. The result was a short, hidden-camera comparison between humans and animals. Newspapers compared the two films and agreed that Haanstra had shown his great ability in both his free film and the commissioned film. Even tied to a commission, his personal vision and poetry remained recognizable. *Delta Phase 1* not only gave the Delta Works a face – after all, the Dutch had also been able to watch the closure of the Veerse Gat on television and cinema newsreels – but also a soul.

Two weeks later, on 4 April, Haanstra screened his film in Veere. At a screening organized especially for the residents of Veere and Arnemuiden, *Delta Phase 1* was shown, along with three other films by Haanstra. The choice of these other films already betrays that the approach was different from the festive screening in The Hague. Besides the Oscar-winner *Glas* (*Glass*; 1958), two older Haanstra films were screened: *. . . En de zee was niet meer* (*. . . And the Sea Was No More*; 1955) and *Strijd zonder einde* (*The Rival World*; 1955). The latter film was a commissioned film for Shell about insect control. Above all *. . . En de zee was niet meer* must have given the show a character of its own.

In *. . . En de zee was niet meer* (title taken from the Book of Revelation), Haanstra films the decline of the towns on the former Zuiderzee. The film was commissioned by the Ministry of Education, Arts and Sciences and aimed to capture the folklore of the towns around the new IJsselmeer. The film begins light-hearted and playful, showing the cycle of life and work of the people dressed in traditional costumes, in which the sea and fishing played an important role. But now the sea is no more, closed off by the Afsluitdijk and partly drained. The centuries-old, traditional way of life is coming to an end. The cycle of life continues and, after baptism, youth and marriage, we have now come to the farewell: a funeral procession accompanies the dead to their final resting place. After a handful of earth is thrown onto the coffin, Haanstra cuts to a crane dumping clay into a nearly closed sea hole. The sea gives way to a new land. The film ends with the skeleton of an old fishing boat being unearthed on the newly raised land – like a fossil from the past.

For the inhabitants of Veere and Arnemuiden, this was a clear reference. That afternoon in Veere, the atmosphere might have been not as cheerful and optimistic as during the screening in The Hague, the seat of the national government.

Introducing the Henri Lefebvre's Rhythmanalytical Approach

Guido Borelli

This essay explores Henri Lefebvre's rhythmanalytical approach, focusing on the way in which it bridges the relationship between time, space and everyday life. Central to Lefebvre's thought is the notion that physical space gains reality through the energy and rhythms deployed within it, which rhythmanalysis seeks to understand. The essay traces Lefebvre's influences, including those of Gaston Bachelard and Georges Gurvitch, and examines how rhythmanalysis critiques the dominance of linear time in capitalist societies, while emphasizing the body's role in perceiving and shaping rhythms of life. Through an analysis of Lefebvre's *Éléments de rythmanalyse*, this essay highlights rhythmanalysis as a powerful instrument for critically examining social space and time, offering new and creative insights into the study of the urban within the social and the spatial disciplines.

The Secret Garden of Henri Lefebvre

During his long and eventful life, Henri Lefebvre (Hagetmau 1901 – Navarrenx 1991) wrote more than sixty books on topics such as philosophy, everyday life, sociology (rural at first, later urban), the state and Marxism. In addition, Lefebvre delved into the thoughts of Nietzsche (1939) and the works of several previous French thinkers and writers: Descartes (1947a), Diderot (1949), Rabelais (1955a) and Musset (1955b). He also published three short theatrical works: a volume on the paintings of Édouard Pignon (1956), a novel titled *Le mauvais temps* (1937), written under the name Henriette Valet – his first wife – and a geographic travel book about his beloved Germany (1958), edited by photographer Martin Hurlimann. Amid these pursuits, Lefebvre also found time to focus on Hitler's rise to power, sexology, the Paris Commune of 1871, the movements of May 1968, and the critique of Le Corbusier's urban planning in Pessac. This variety of interests is reflected in his vast output of essays published in journals, spanning more than sixty years.

Despite the heterogeneity of the topics he tackled, this massive body of work is characterized by the French philosopher's particular writing style. What made Lefebvre's style distinctive is not simply that he consistently rejected what, in the pompous language of academia, are called 'bibliographic references', nor that his tone is often highly colloquial (Lefebvre generally did not write by hand but dictated to a typist, with whom he often maintained *histoire tendre*). Lefebvre's writing style is marked by the dissemination of concepts. In his books, Lefebvre would introduce, often abruptly, thoughts, notions and ideas, only to pick them up again hundreds of pages later or, quite often, in other works. Sociologist Harvey Molotch (1993), in a review of the English edition of *La production de l'espace*, although generous in his praise for Lefebvre's text, could not refrain from commenting: 'Lefebvre writes densely, with digressions and gratuitous allusions to other scholars, with mysterious phrases and an inconsistent and disorganized development of content. Lefebvre writes terribly!'

In this writing style, the concept of rhythm – which the French philosopher, by the end of his long life, systematized into the method of rhythmanalysis – is certainly one of those recurring ideas. Rhythmanalysis, understood as a research methodology dedicated to understanding the links between time, space and everyday life, explicitly appeared in Lefebvre's work only in his later years, when the French philosopher was well into his eighties. In 1985, in an essay titled 'Le project rythmanalitique', Lefebvre explicitly anticipated the content of a later publication, *Éléments de rythmanalyse* (1992), published posthumously. However, it should be noted that the concept of 'rhythm' appears, explicitly or latently, in almost all of Lefebvre's writings. As early as his first essay, published in 1925 in the Parisian Surrealist-inspired journal *Philosophies* and titled 'Positions d'attaque et de défense du nouveau mysticisme', the 24-year-old Lefebvre, freshly graduated in philosophy from the Sorbonne, criticized idealism for 'holding the world at bay'. He argued that existence is a fact with which it is impossible not to engage, and claimed that life is an adventure that must be lived fully.

Lefebvre's focus on rhythms began to take shape amid his interest in everyday life. In the second volume of *La critique de la vie quotidienne* (1961), the French philosopher used the term 'rhythmology', announcing the need for a critical methodology for the study of social time in capitalist industrial society through the conflict between linear and cyclical forms of time. A reference to rhythmanalysis is also present in *Toward an Architecture of Enjoyment*, a manuscript written in 1973 for his Spanish former student and friend Mario Gaviria, but not published until 2014. In the third volume of *La critique de la vie quotidienne* (1981, 129), in a paragraph titled 'Space and Time', Lefebvre famously states: 'when Marxists have dealt with rhythms, they have considered them solely on the basis of labour'.

However, it is in his magnum opus on space, *La production de l'espace* (1974), that Lefebvre lays the foundations for a grand project that combines the relationships between space and time through the body – or, more precisely, through the everyday life of the body. In *La production*, Lefebvre begins to consider that the study of rhythms, owing to its greater concreteness, effectiveness and proximity to appropriation (understood as the appropriation of the body as a spatial practice), could complete or even replace psychoanalysis. In *La production* (1991, 31), Lefebvre tersely asserts – almost as an aphorism – that 'every society – and therefore every mode of production with its subvariants – produces a space, its own space'. Therefore, to understand this relationship, it is necessary to conduct 'a study of this space that is able to grasp it as such, in its genesis and form, with its time or specific times (the rhythm of everyday life), and its particular centres and polycentrisms (agora, temple, stadium etc.)' (*ibid.*). Given that, for Lefebvre, space is a social product, a theory of the practices necessary to grasp its complexity must necessarily begin with the study of social rhythms. In this long but essential quotation, Lefebvre (*ibid.*, 87) effectively explains the matter using the following metaphor:

> Social spaces interpenetrate one another and overlay themselves upon one another. They are not things, which have mutually limiting boundaries and which collide because of their contours or as a result of inertia. . . . Nor can such spaces be considered empty 'mediums', in the sense of containers dis-

> tinct from their contents. . . . A much more fruitful analogy, it seems to me, may be found in hydrodynamics, where the principle of the superimposition of small movements teaches us the importance of the roles played by scale, dimension, and rhythm. Great movements, vast rhythms, immense waves – these all collide and 'interfere' with one another; lesser movements, on the other hand, interpenetrate. If we were to follow this model, we would say that any social locus could only be properly understood by taking two kinds of determinations into account: on the one hand, that locus would be mobilized, carried forward, and sometimes smashed apart by major tendencies, those tendencies which 'interfere' with one another; on the other hand, it would be penetrated by, and shot through with, the weaker tendencies characteristic of networks and pathways.

Following Lefebvre, the explanation (along with the method for arriving at the explanation) would come almost twenty years later, at the end of his life, with the publication of *Éléments de rhythmanalysis*. René Lourau (1992), a student and friend of Lefebvre, in his affectionate introduction to the posthumous edition of the book, observed:

> Rhythmanalysis becomes a supporting discipline for the critique of everyday life. . . . It is situated at the boundary between the physical, the psychological, the social, and at the heart of the everyday. . . . The rhythmanalytic project remained, until the end, Henri Lefebvre's secret garden . . . a garden that remained closed for a long time – 'jardin cerrado para muchos' (Lorca). (*ibid*., 9, translated by author).

Lourau's introduction offers a stimulating and clarifying analogy between Lefebvre's hydrodynamic metaphor and the first chapter of Italo Calvino's book *Palomar* (1983), titled 'Reading a Wave'. For Lourau (1992, 8), all the complexities of rhythmanalysis are encapsulated as follows:

> One cannot observe a wave without taking into account the complex aspects that contribute to its formation and the equally complex aspects to which it gives rise. These aspects vary continually, so that one wave is always different from another wave; but it is also true that every wave is the same as another wave, even if not immediately contiguous or successive; in short, there are forms and sequences that are repeated, albeit irregularly distributed in space and time.

The Genealogy of Rhythmanalysis: Duration, Social Time, Moments and Le Soleil Crucifié

If Lefebvre's intention to analyze rhythms was the result of a long maturation, then some bibliographic references are necessary to establish its genealogy. The first source of inspiration was undoubtedly the work of the French philosopher Gaston Bachelard. In fact, the term rhythmanalysis (*rythmanalyse*) was not invented by Lefebvre – nor by Bachelard. The first to use it was the Portuguese experimental psychologist Lúcio Alberto Pinheiro dos Santos, who sought refuge in Brazil in 1927 to escape the dictatorship of general Carmona. In 1931, Pinheiro dos Santos wrote an essay titled 'A ritmanalise' for a non-existent journal published by the equally imaginary Sociedade de Psicologia e Filosofia of Rio de Janeiro. The essay would surely have fallen into oblivion if the author had not sent it to Bachelard, who was enthusiastic about it and used it extensively in the final chapter of his book *La dialectique de la durée* (1936). However, as Baptista (2010) reports, the manuscript by Pinheiro dos Santos has never been found, so what remains is Bachelard's use of the concept of rhythmanalysis.

In *La dialectique*, Bachelard aimed to understand the complexity of life through a critique of Bergson's concept of duration. For Bachelard, life consists of a plurality of durations, each with its own rhythm, connections and relations with other durations. Since each of these pluralities owes its stability to complex forms of agreement (or, sometimes, discordance) between an individual's material, biological or psychological rhythms, rhythmanalysis presents itself as a (new) component of the psychological sciences on par with psychoanalysis. Its purpose is therefore to 'heal the suffering soul – especially the soul that suffers from time, from spleen' (Bachelard 1936, 12). For Bachelard, through a rhythmic life, it is possible to free the soul from false permanences and ill-formed durations by reorganizing it in time and space. Rhythmanalysis 'seeks opportunities for rhythm everywhere. It is convinced that natural rhythms correspond or can easily overlap one another. In this way, it warns us of the danger of living in the wrong moment, signalling the fundamental need for temporal dialectics' (*ibid.*, 143). The rhythmanalysis proposed by Bachelard focuses on the analysis of the 'most appropriate' rhythms to achieve a mental-emotional state capable of animating life through the lightness of intellectual freedom. It is reassuring, often poetic: rest as a 'right of thought' can be enlightened, spiritualized and poetized by living well-regulated temporal diversities.

Lefebvre read Bachelard as a romantic who became a scientist or sociologist in order to interpret the forms of alienation that erode the poetics of ordinary people. This interpretation of poetic experiences in everyday life brings Lefebvre very close to Bachelard's thinking. What distinguishes him is the maintenance of a critical focus that links poetic issues to the scientific ones of critical theory (Revol 2015, 214-215).

The second source of inspiration Lefebvre drew from his mentor, the Russian-French sociologist Georges Gurvitch. For Gurvitch, society as a temporal reality is both an object and a subject in constant becoming; it is the product of all the combined consciousnesses and memories that emerge from the infinite intersection of instances to be remembered and anticipated. On this basis, he founded a 'dynamic' sociology that opposed the 'static' one, which too often mystified sociological consciousness and ignored the totality of progress. For Gurvitch, time is always at work in social life, but in various ways and following different rhythms: the times of life and death, knowledge and ignorance, recognition and denial, celebration and oblivion. From Gurvitch (1958), Lefebvre borrowed the definition of 'social time', which was in direct opposition to the classical idea of a singular, homogeneous time, conceived as an *a priori* of collective life. This helped him further clarify both the belief that time is not merely mental, but also social, biological, physical, cosmic, linear and cyclical, as well as the dialectical relationship between linear and cyclical time, which constitutes one of the fundamental pillars of the rhythmanalytic method.

The third source of inspiration Lefebvre derived from the theory of moments, which he began to develop in the early 1920s. For Lefebvre, moments are those states of intense experience in everyday life that offer the possibility of a critique of the everyday itself. These are experiences related to strong sensations of disgust, intense pleasure, panic, joy and so on. Moments generate the premises for a different everyday life and simultaneously break the continuum of the present. Lefebvre describes them thus in his autobiography, *La somme et le reste* (1959):

> A moment relates to the whole of life; it includes all the times lived . . . but nevertheless, it is a partial fragment of it. It is a slow journey that leads to a brightness, a shock, a semi-understanding after which what was hitherto merely perceived or understood becomes a theme for reflection, without however exhausting its own fruitfulness. The moment has its own memory, its own formation and maturation; it condenses around a central image that exists, but disappears in spontaneous life. . . . This event retrospectively sheds light on a large part of reality, and [at the same time] on the time that leads up to it. It surpasses it. It prolongs itself indefinitely. This takes time, a lot of time: time [for] the whole cannot be kept (Lefebvre 1959, 246, translated by the author).

Lefebvre traced his awareness of the existence of moments to a movement of rebellion and horror against the bigoted religion that poisoned Navarrenx, the town of his maternal family, where young Henri spent his summer vacations, cultivating a deep disgust and disdain for 'a village of merchants and artisans devoid of charm' (Lefebvre 1947b, 229). Around the years 1919-20, he realized that, besides post-adolescent malaise, the conformism and moral order imposed by the emerging Catholic bourgeoisie in the immediate postwar period were valid reasons to justify revolt. To represent the irreplaceable need for this revolutionary impulse, Lefebvre used a mythological image: *le soleil crucifié* (the crucified sun). During his long summer walks in the Vallée du Gave d'Oloron, he had been intrigued by the remarkable crucifixes placed at crossroads, which bore a disc at the intersection of the arms. Lefebvre had at first noticed the uniqueness of these crosses without giving them much importance. Indeed, the young philosopher remembered having heard or read somewhere that these discs represented the instruments of Christ's passion: the crown of thorns or the spears of Roman soldiers, inscribed within those circles. One day, he had a sort of *vu jàdé*:

> That day, sitting on the base of the cross, an idea that I had undoubtedly been maturing for a long time seized me. I stood up abruptly and looked at the cross above my head: 'They crucified the sun! They crucified the sun!' I walked away in horror from that place, from that object. But the image that had for the first time appeared clearly to me after I had walked that path many times, had to continue its own journey. . . . The sun nailed to the cross of torment was youth, my youth. . . . I swore to myself to extract the deadly nails, to free the solar prince, to burn the cross of death. But, at the same time, a new curiosity seized me. Who were these men – my ancestors – who completed this act, who created this grand and tragic symbol and placed it – they, obscure peasants – at the crossroads of paths? Why? When? (Lefebvre 1947b, 45, translated by the author).

For Lefebvre, moments – understood as differential experiences in everyday life – enlighten us in understanding the states of alienation into which capitalist society has thrown us. Regarding these states of alienation, he spared no criticism, even against religion (specifically the Catholic faith, which he perceived as bigoted and old-fashioned within his maternal family). The fifth chapter of the first volume of *La critique de la vie quotidienne* (1947b) is titled 'Notes Written on a Sunday in the French Countryside' and is a diary of everyday life in a small village in Béarn, the region from which Lefebvre originated, in the immediate postwar period. This chapter is often cited by scholars due to Lefebvre's fierce reckoning with the Catholic Church, which, like economic fetishism, is guilty of producing alienation in the lives of humble believers: 'Its power comes from the fact that it penetrates everyday life, . . . a psychological and moral technique of extreme finesse and precision' (*ibid.*, 260). Lefebvre adopted from Marx's *Economic and Philosophical Manuscripts* (1848) the concept of alienation as central to the critique of everyday life. We can assume that if the theory of moments represented for Lefebvre the metaphysical foundation for the critique of everyday life, then the theory of rhythms completes it by offering itself as its scientific-methodological foundation.

The Body, Time and Space

Rhythmanalysis emerges in Lefebvre's thought as a necessary concept to resolve the interplay between linear and cyclical rhythms generated by the capitalist mode of production. Linear time has become hegemonic in capitalism because, according to Lefebvre, it has been transformed into an abstract entity, needed solely to measure the exchange value inherent in the production process: 'time is money' (Moore 2013, 73). Thus, capitalism's linear time promises innovation and progress, but offers only monotony and boredom in return. On this point, Lefebvre makes a crucial distinction between rhythm and repetition. Cyclical rhythms preserve differences within their recurring sequences; a sunrise or sunset is always unique, even though they occur every day. Rhythms do not erase the possibility of desire and discovery; hunger and thirst appear with ever-changing nuances. Linear rhythms, on the other hand, eliminate all differences in an attempt to render time homogeneous, equivalent and interchangeable. The formal and material identity of a work cycle is exactly recognizable, repetitive, identical to itself and equally generative of exhaustion and boredom. The task of rhythmanalysis in modern capitalist society is to equip a critical tool, a methodology, to analyze the dominance of linear time over cyclical time.

For Lefebvre, despite the hegemony of linear time, there are always forces – more or less residual – that oppose capital's ability to completely colonize everyday life. Even though the qualitative has virtually disappeared in favour of the quantitative, Lefebvre insists that it is from this virtuality that we must start, by establishing a dialectical relationship between the linear and the cyclical, between power and resistance. However, a dialectical relationship requires three terms, as dualism is not sufficient, because two terms create a contradiction rather than a relationship. For this reason, Lefebvre believes that the rhythmanalysis of the repetitions characteristic of linear time and the differences inherent in cyclical time must be integrated by introducing a third element: the body, so that the relationship progresses from the abstract to the concrete.

At this point, because a body is not normally able to perceive rhythms that are outside its own perceptual scale (for example, infrared and ultraviolet radiation or ultrasound and infrasound, borrowed from physics), a method is needed to successfully undertake an analysis of rhythms. In this regard, *Élements de rythmanalyse* (1992) aims to guide the reader in rhythmanalytical praxis.

'Vue de la fenêtre'

In chapter three of *Élements de rythmanalyse*, titled 'Vue de la fenêtre' ('View from the Window'), Lefebvre offers us a demonstration of rhythmanalytical practice. From the windows of his apartment on Rue Rambuteau 30, in the fourth arrondissement of Paris, Lefebvre observed the daily comings and goings and the rhythms produced by the newly built Beaubourg-Centre Georges Pompidou. On one hand, he revived the tradition of some French Impressionist painters, particularly Camille Pissarro and Gustave Caillebotte (Durand-Ruel Snollaerts 2017), who effectively used the near/far viewpoint offered by a window overlooking the street (*fenêtre sur la ville*) in their works. On the other hand, he goes beyond mere representation of an observation point because 'the window responds'. Rhythmanalysis, Lefebvre asserts, is not a contemplative or merely descriptive or allusive activity; it is a critical activity.

> No camera, no image or series of images can show these rhythms. It requires equally attentive eyes and ears, a head and a memory and a heart. A memory? Yes, in order to grasp this present otherwise than in an instantaneous moment, to restore it in its moments, in the movement of diverse rhythms (Lefebvre 1992, 36).

The window suggests to Lefebvre various hypotheses – the crowd and its wandering – used to establish the dialectic of rhythms in time and space, but also the forms and gestures that the power of the state has inscribed in that place:

> The bodies (human, living, plus those of a few dogs) that move about down there, in the car-wrecked swarming whole, impose a law? Which one? An order of grandeur. The windows, doors, streets and façades are measured in proportion to human size. The hands that move about, the limbs, do not amount to signs, even though they throw out multiple messages. But is there a relation between these physical flows of movements and gestures and the culture that shows it-

self (and yells) in the enormous murmur of the junction? The little bistros on the rue R., the boutiques, are on a human scale, like the passers-by. Opposite, the constructions wanted to transcend this scale, to leave known dimensions and also all models past and possible behind; leading to the exhibition of metal and frozen guts, in the form of solidified piping, and the harshest reflections. And it's a meteorite fallen from another planet, where technocracy reigns untrammelled. . . . Does the state-political order write across this scene, with the signature of the author? Without doubt, but the time and the age that inscribe themselves in the performance of this spectacle, that give it meaning, should not be forgotten (*ibid.*, 33-34).

For Lefebvre, the window overlooking Rue Rambouteau is not a mental space from which one's gaze can drift into abstract interpretations, but rather a 'practical, private and concrete' place that offers views far beyond 'spectacles or perspectives prolonged mentally'. Here, the French philosopher focuses on the socio-economic organization of urban-mercantile society to reaffirm that, from a Marxist perspective, commodities take over everything, and, from a Lefebvrian perspective, (social) space and (social) time become the space and time of markets and exchange. In this way, extending the concept of commodity fetishism found in Marx, space and time in Rue Rambouteau become things that – this is the crucial point – contain rhythms within them.

The necessity of perceiving the rhythms of things and people necessarily involves perceiving one's own rhythms through one's body: the chapter 'Vue de la fenêtre' must therefore be considered both as the moment when the rhythmanalyst encounters others and their rhythms, and as the method for deciphering those rhythms. Unlike phenomenological analysis, rhythmanalysis does not merely consider that 'there are rhythms', but also that rhythms are produced within a given situation. Its goal is to qualify these rhythms and illustrate the causes and effects of the interaction between time, space and energy produced by bodies and things in those times and spaces. For the rhythmanalyst, this interaction is dialectical and forms the starting point of the critical methodology of the processes of reification.

The references to the political order reigning in Rue Rambouteau and to the omnipresent state with its absolute technocracy, 'which cannot be seen from the window but is present in this present' (Lefebvre, 1992) show us how Lefebvre's analysis shifts – or more aptly 'wanders' – between the observation of street rhythms and critique of society. The succession of these alternations, repetitions and differences suggests to the rhythmanalyst that this present is ordered by something coming from elsewhere, which is not visible from the window. To capture this presence, the rhythmanalyst must use their body 'like a metronome' (*ibid*., 19 and passim):

> In order to grasp this fleeting object, which is not exactly an object, it is therefore necessary to situate oneself simultaneously inside and outside. . . . However, to grasp a rhythm, it is necessary to have been grasped by it; one must let oneself go, give oneself over, abandon oneself to its duration (*ibid*., 27).

Empiricism and Substance vs. Abstraction and Ideology

It is interesting to note how Lefebvre felt the need to bid farewell to a century lived almost entirely theoretically through a methodological project. It seems as though he wanted to conclude his intellectual journey by refuting the widespread pluralist criticism that Marxist approaches tend to confuse ideology with substance and abstraction with empiricism (McLennan 1989). To neutralize this critique, Lefebvre turned to Marx (1844) and the Third Manuscript, 'Private Property and Communism', from which he adopted the idea that 'the senses have therefore become theoreticians in their immediate praxis'. Thus, on the one hand, the (empirical) study of the body's rhythms and its subjugation to training and social rules (see the fourth chapter of *Élements de rythmanalyse*, titled 'Dressage') is indispensable not only for analyzing how capitalism shapes classes but also, more importantly, for understanding how it acts as a system based on the alienation of bodies and their lifetimes and spaces. To clarify this point, Lefebvre (1992, 33) proposes a fundamental observation for the practice of rhythmanalysis as a research method: 'the implication in the spectacle entails the explication of this spectacle'.

The concept of implication is of fundamental importance for the practice of rhythmanalysis. Lourau (1992) did not miss this point by Lefebvre, noting how it introduces important questions regarding the use of the body of the researcher-rhythmanalyst in rhythmanalytical practice. Specifically, Lourau carefully reflects on what it means for a researcher to 'be implicated' in their object of research. According to Lourau, this involves 'an endless, incomplete, uncertain work that cannot overcome the indeterminacy of the situation'. The result is not 'explanation' as in classical logic or classical sociology. The analysis of involvement 'explains' more or less our place 'within the spectacle' of the world, and in doing so provides some illumination of the world – or its representation. For Lourau, recognizing these implications means doing the bulk of the social and analytical work, as the obstacles to such recognition constitute the main part of the field of analysis. The concept of implication is particularly useful for understanding the practice of rhythmanalysis, because implications are not just 'responses to' but especially 'relations to'. For Lourau (1969, 12), our bodies are akin to those of ventriloquists who speak 'only because the institution speaks through us, because we literally have it 'under our skin'. Everyday life is entirely composed of an institutional fabric that involves a certain degree of consent, adhesion, commitment and participation (if not integration), and we are all involved with the institution, even when we do not objectively belong to its established form. Lourau offers an effective example to support this thesis:

> It does not matter whether I am rich or poor, whether I am an employee or a shareholder: the institution of the bank, and the institutions connected to it, concern me. I deal with the bank, even if I have no bank account, no savings book, no postal account, because the bank is the established form of regulation concerning the circulation and accumulation of capital (Lourau 1969, 10, translated by the author).

Through the researcher's implications, rhythmanalysis recognizes the importance of arbitrariness and suggests methodological freedom rather than an obstinate pursuit of positivist truths through poorly designed research. Arbitrariness highlights the methodologically nomadic nature of rhythmanalysis, which refuses to be confined within disciplinary boundaries that would limit its trajectories. This does not mean that rhythmanalysis is unresponsive to methodological discussion. Rather, it means that it is based on the understanding of social experiences as reverberating and impregnated, and that it situates itself methodologically within social phenomena, in contrast to research practices that position themselves above social phenomena.

Everything that established research methodologies consider fundamental requirements (the claim of objectivity in analysis, the non-partisanship of the researcher), and upon which social sciences have drawn legitimacy over the last century and a half, represent for Lefebvre a fatal limitation that often leads the researcher (and their research) to the famous Hegelian aphorism: 'what is known is not necessarily understood'. With rhythmanalysis, Lefebvre returns for the last time to the project he cultivated throughout his life: the critique of everyday life. Whereas in established social sciences, everyday life is a descriptive and analytical concept – taken 'for granted' in any case; in Lefebvre's project this conception is insufficient because it does not define the rhythmic essence of the desires and needs of everyday life as rhythms interacting in space. It does not capture the needs, desires, unconscious reflections and passions.

Conclusion

Élements de rythmanalyse shares the same destiny as most Lefebvrian reissues of the last thirty years: receiving attention and (re) opening the scientific debate only after being translated into English. By being translated, *Élements de rythmanalyse* provides further possibilities for empirical methodologies in applied research. This is evidenced by the proliferation of studies that since 2004, the year in which the work was made accessible to Anglo-Saxon scholars in their own language, have referred in various ways to rhythmanalysis as a research method. In this regard, it should be considered that, as the elaboration of rhythms is not presented by the author himself as a closed system, but only sketched out in this small treatise, many researchers have devoted themselves to not only exploring its empirical potential but also to trying to define its procedural aspects in some way. The Lefebvrian concepts of 'polyrhythmia', 'eurhythmia' and 'arrhythmia' have been variously declined to show that places and times are constituted and function through mobile flows of capital, objects, energy, or matter passing through and around them. In this respect, places and times exist in a constellation of rhythmic entanglements: polyrhythms that contain multiple temporalities in harmony (eurhythmia) or in contrast (arrhythmia) with each other in an ever-evolving rhythmic assemblage.

According to Chen (2018, 6), at the conceptual level, there are two possibilities for practising rhythmicity. The first approach is inductive: it focuses on a distinct rhythm and weaves the multiplicity of rhythmic assemblages together by linking the different points of rhythmic production. Starting from a distinct rhythm, the rhythmanalyst traces the polyrhythms of social processes by exploring the reciprocal influence between a singular rhythm and the totality of other rhythms. The second approach is deductive, in which the plurality of practices of temporal spacing, alignment, negotiation and realignment of rhythmic assemblages present at a given point influences the units of rhythms within the polyrhythmic whole and facilitates the rhythmanalytic recognition of singular rhythms.

Finally, in full coherence with the Anglo-Saxon analytical tradition, a handbook was published (Lyon, 2019) dedicated to the practice of rhythmanalysis. The handbook complements more traditional methodological guides to qualitative interviewing, longitudinal data analysis, community studies, etc. Lyon's work, in addition to formally inaugurating the entry of rhythmanalysis within the 'established' research methodologies, reports an interesting state of the art that testifies to the breadth and heterogeneity of research interests aroused by Lefebvre's method over the last fifteen years.

The number of studies that utilize rhythmanalysis in original ways is continuously growing, and it is not our aim to offer a review of them here. These are generally works that deal directly with urban spatiotemporal themes, as well as those that explore technical possibilities (audiovisuals, timelapse, sound recording, drone use, etc.), kinetic modalities (walking, traveling by train, bus, metro, ferry, car, motorcycle etc.) and static ones (the window on Rue Rambuteau, the table at the bistro, the bench etc.).

This is certainly a positive sign. The fact that there is a widespread, heterogeneous use of Lefebvrian rhythmanalysis by scholars all over the world – and, as mentioned in the previous section, 'differently implicated' – undermines the position of those who shy away from any Marxist imprint in their research activities. The extraordinary reception of a research practice developed by a Marxist scholar (albeit a very unorthodox one, as Lefebvre certainly was) attests to the possibility of conducting original research using methodologies rooted in critical ideology, while at the same time leaving 'space' for the researcher's own implication. The hope is that this will contribute to bringing a breath of fresh air into the immense realm of research dealing with time, space and energies in the urban context, and to focusing research topics on the conditions and possibilities for each of us to approach the study of the everyday in a less tedious and serious manner, with greater attention paid to the minutiae of the ordinary.

In the hope of attracting renewed attention to everyday ordinariness, capable of highlighting the possibilities of a disalienated existence, I could not find a more fitting conclusion here than that offered by Georges Perec (1989, 9) to stimulate not the discovery of new fields of research, but rather the creation of new ways of researching what we think we already know, but that, paraphrasing Hegel once again, 'is not necessarily understood':

> Newspapers talk about everything, except the daily. Newspapers bore me; they teach me nothing; what they recount does not concern me, does not question me, and does not answer the questions I ask or would like to ask. What really happens, what we live, the rest, all the rest, where is it? What happens every day and comes back every day, the banal, the everyday, the obvious, the common, the ordinary, the infra-ordinary, the background noise, the habitual – how can we account for it, how can we question it, how can we describe it?

References

Bachelard, G. (1936). *La dialectique de la durée* (Paris: Boivin).
Baptista, P. (2010). *O filosofo fantasma: Lucio Pinheiro dos Santos* (Lisbon: Zéfiro).
Calvino, I., (1983). *Palomar* (Turin: Einaudi).
Chen, Y. (2018). *Practising Rhythmanalysis: Theories and Methodologies* (Lanham: Rowman & Littlefield).
Durand-Ruel Snollaerts, C. (2017). *Camille Pissaro: Fenêtre sur la ville* (Rouen: Éditions des Falaises).
Gurvitch, G. (1958). *The Spectrum of Social Time* (Dordrecht: D. Reidel).
Lefebvre, H. (1925). 'Positions d'attaque et de défense du nouveau mysticism', *Philosophies, 5-6*, 471-506.
Lefebvre, H. (1939). *Nietzsche* (Paris: Éditions sociales internationales).
Lefebvre, H. (1947a). *Descartes* (Paris: Éditions hier et aujourd'hui).
Lefebvre, H. (1947b). *Critique de la vie quotidienne, Vol. I* (Paris: L'Arche).
Lefebvre, H. (1949). *Diderot* (Paris: Éditions hier et aujourd'hui).
Lefebvre, H. (1955a). *Rabelais* (Paris: Éditions hier et aujourd'hui).
Lefebvre, H. (1955b), *Musset* (Paris: L'Arche).
Lefebvre, H. (1956). *Pignon* (Paris: Le musée de poche).
Lefebvre, H. (1959). *La somme et le reste* (Paris: Nouvelles Éditions Françaises).
Lefebvre, H. (1961). *Critique de la vie quotidienne, vol. II* (Paris: L'Arche).
Lefebvre, H. (1991 [1974]). *The Production of Space* (trans. D. Nicholson-Smith). (New York: John Wiley and Sons Ltd).
Lefebvre, H. (1981). *Critique de la vie quotidienne, vol. III* (Paris: L'Arche).
Lefebvre, H. (1992). *E´lements de rythmanalyse* (Paris: Persée).
Lefebvre, H. (2004). *Rhythmanalysis: Space, Time and Everyday Life* (trans. S. Elden & G. Moore). (London: Continuum).
Lefebvre, H.; ed. L. Stanek (2014). *Toward an Architecture of Enjoyment* (Minneapolis, MN: University of Minnesota Press).
Lefebvre, H., & Hurlimann, M. (1958). *Germany* (Mulhouse: Braun).
Lefebvre, H., & Régulier, C. (1985). 'Le project rythmanalitique', *Communications, 41*, 191-199.
Lourau, R. (1969). *L'instituant contre l'institué: essais d'analyse institutionnelle* (Paris: Anthropos).
Lourau, R. (1992). 'Henrisques', in H. Lefebvre, *E´lements de rythmanalyse*, 5-10 (Paris: Persée).
Lyon, D. (2019). *What is rhythmanalysis?* (London: Bloomsbury).
Marx, K. (1844). *Ökonomisch-philosophische Manuskripte aus dem Jahre 1844*. Available: https://www.marxists.org/archive/marx/works/1844/epm/index.htm (accessed 18 June 2025).
McLennan, G. (1989). *Marxism, Pluralism, and Beyond: Classic Debates and New Departures* (London: Polity Press).
Molotch, H. (1993). 'The Space of Lefebvre', *Theory and Society, 22*(6), 887-895.
Moore, R. (2013). 'The Beat of the City: Lefebvre and Rhythmanalysis', *Situations, 5*(1), 61-77.
Perec, G. (1989). *L'infra-ordinaire* (Paris: Éditions de Seuil).
Revol, C. (2015). *La rythmanalyse chez Henri Lefebvre (1901-1991): Contribution a` une poe´tique* (PhD thesis, Universite´ Jean Moulin Lyon 3).
Valet, H. (1937). *Le mauvais temps* (Paris: Grasset).

Decoding Socio-Environmental Processes through Rhythm-analysis

Pınar Şefkatli

This paper outlines the technique of rhythmanalysis as a transdisciplinary method for understanding socio-environmental processes in urban settings. Drawing on Lefebvre's, Gandy's and Adam's theoretical framework, I explore three rhythms that shape urban environments, to propose new perspectives on sustainability and climate change: social rhythms, ecological rhythms and infrastructural rhythms. These rhythms vary from daily human activities to seasonal biodiversity cycles, shaping and reshaping each other in the urban environment. In this context, I propose that combining artistic and scientific research approaches informs a better understanding of how these rhythms manifest, as well as how they harmonize and clash with each other, enabling both a rigorous data collection and representation and a possibility to tune with sensorial experiences. In this light, rhythmanalysis opens a window into research frameworks that transcend disciplinary boundaries and bridge between arts and sciences, fostering an inclusive inquiry towards cultivating sustainable urban futures.

Introduction

Urban environments are characterized by the interplay of natural cycles, daily human activities and interactions, and the infrastructural systems that support everyday life. Within this complex web, rhythms shape how time and space unfold, ranging from the cyclical patterns of day and night to the routines and habits of individuals and communities, as well as spatio-temporal patterns of urban supply systems, grids and networks. Although a wide range of studies recognizes these rhythms and their implications on urban life and processes (Brighenti & Kärrholm, 2018; Matos Wunderlich, 2013; Nash, 2020), understanding how they relate to each other and how they converge and interact in cities has not been fully explored in twentieth-century urban social theory (Heynen et al., 2006). Can this gap in research and practice create opportunities to integrate arts and sciences, developing viewpoints that transcend traditional disciplinary boundaries?

Rhythmanalysis offers a powerful lens for understanding how various dynamics of ecological cycles, social interactions and infrastructural systems intersect in urban environments (Lefebvre, 2013). Studies of rhythms in the urban context emphasize not only the spatial qualities of these dynamics but also their temporal aspects, exploring how they are repeated and reproduced and how they relate to or diverge from each other over time (Adam, 2005). This approach enables us to position socio-environmental and infrastructural conditions within their spatio-temporal contexts (Gandy, 2014) by combining various qualitative (experience-based) and quantitative (data-based) forms of knowledge. Representing these rhythms visually serves as a 'boundary object', helping to create a common ground in which the paradigms of arts and science can engage in transdisciplinary settings (Nevejan & Sefkatli, 2020). But what is rhythmanalysis? How does it enable us to understand these intermingled dynamics?

Understanding urban biodiversity and climate change benefits a multi-scalar perspective that bridges the arts and sciences, framing the climate as a 'hybrid entity' shaped by both environmental and biophysical elements as well as socio-cultural knowledge, affect and practice (Popke, 2016). Lefebvre reminds us that these challenges are not isolated phenomena but 'manifestations of the more "global problem" of capitalism's survival' through its reproduction of social and spatial relations (Butler, 2023). To understand these perspectives, we need a 'multi-scalar and multi-species framework' that uncovers ecological complexities within urban environments (Gandy, 2023). By approaching socio-environmental challenges as triggers of 'spatial displacement and temporal disorientation' (Gandy, 2023), we can illuminate their associations with cultural landscapes through scientific and artistic methods. This interplay of space and timescales sets the foundation for exploring rhythmanalysis, revealing opportunities to harmonize social, ecological and infrastructural systems in the face of climate crises (Figure 1).

In this article, I will explore the theoretical foundations of rhythmanalysis in socio-environmental contexts, drawing on the works of Henri Lefebvre, Matthew Gandy and Barbara Adam. I will then discuss the potential of rhythmanalysis as a methodology for identifying, documenting and analyzing social, ecological and infrastructural rhythms and their interrelationships through artistic and scientific research methods. Finally, I will argue the relevance of rhythmanalysis in bridging these two fields, revealing diverse temporalities ranging from immediate, daily rhythms to long-term environmental cycles. By bringing to light how such rhythms merge and unfold in the urban environment and the elements that structure them through arts and science, we can propose innovative solutions that render them more attuned to one another. Ultimately, rhythmanalysis emerges as a powerful resource for shaping sustainable and resilient urban futures, inviting a collaborative discourse between arts and sciences.

Theoretical Reflections on Social, Ecological and Infrastructural Rhythms

How do rhythms shape urban environments and socio-environmental processes? How can we identify social, ecological and infrastructural rhythms and explore their intricate relationships? What methodologies can we adopt to deepen our understanding of these rhythms through engaging arts and sciences to foster more sustainable and resilient urban landscapes? Let us delve into these inquiries through an exploration of the theoretical frameworks proposed by Henri Lefebvre, Matthew Gandy and Barbara Adam.

Henri Lefebvre and Social Rhythms

In his works *The Critique of Everyday Life* and *Production of Space*, Henri Lefebvre invites us to explore the rhythms that shape both ecological processes and our social lives, highlighting the 'nature-society dialectic of our time' (Foster et al., 2020). At the heart of rhythmanalysis lies the investigation into how natural and social rhythms intertwine. Lefebvre distinguishes these two by identifying two types of rhythms: cyclical and linear (Lefebvre, 1996). Cyclical rhythms arise from natural cycles and processes, which Lefebvre illustrates with examples such as the alternation of day and night, changing seasons and even human biological rhythms like our heartbeats (Lefebvre, 1996). These rhythms are characterized by regeneration and renewal, fostering a sense of continuity and stability within ever-changing urban life. In contrast, linear rhythms reflect human activities and actions, such as work schedules, transportation systems and economic processes, which, according to Lefebvre, are shaped by the imposition of social and political structures. In his critique of daily urban processes, Lefebvre observes how linear rhythms are often normative, prioritizing productivity and efficiency over natural temporalities, reducing time to mere measurable units (Lefebvre, 2013).

Lefebvre's analysis unveils the complex interactions between these two distinct rhythms within our urban environments. He argues that modern cities favour linear rhythms, which often trap us in a cycle of standardized schedules, repetitive timeframes and the relentless 24/7 operations of urban infrastructures (Lefebvre, 2014). These temporal orders can distance urban dwellers from ecological rhythms such as 'seasonal habits of the tides, the weather, day and night' (Jones, 2010). Lefebvre suggests that by recognizing these dynamics of disruption, we can begin to rethink urban life itself and how the rhythms of nature can be harmonized with the rhythms of social life (Crang, 2001). What if, instead of resisting the natural rhythms, urban life embraced them?

Matthew Gandy and Infrastructural Rhythms

Matthew Gandy's work invites us to explore the intricate rhythms generated by urban infrastructures and their crucial role in shaping the socio-environmental processes in the urban context. Urban infrastructures are 'not only material manifestations of political power but also systems of representation that lend urban space its cultural meaning', transforming not just the urban space but also the experiences of those who inhabit them (Gandy, 2005, 39).

Various urban infrastructures – water management systems, energy distribution networks and waste disposal mechanisms – impose their own structured temporalities on natural flows, creating unique rhythms within the urban fabric (Gandy, 2022). For example, water infrastructures like dams and drainage systems dramatically alter the natural patterns of rivers and rainfall, impacting seasonal flooding and the delicate ecosystems that rely on these cycles. By reshaping these natural flows, infrastructure produces new urban rhythms that both accommodate and challenge ecological processes. This dynamic creates 'tensions' between the ongoing maintenance of these systems and their long-term endurance, while impacting the 'specificities and temporal dynamics of micro ecologies in the urban environment' (Gandy, 2022).

Echoing Lefebvre, Gandy embarks on a journey to synthesize various temporalities within an 'integrated analytical framework' (Gandy, 2022). This approach underscores the critical interplay between infrastructure and the socioecological rhythms within our cities. For Gandy, understanding how urban infrastructures – which he calls the 'space-time geographies of modernity' (Gandy,

2022) – shape and interact with socio-ecological processes is essential to developing novel approaches for urban sustainability and resilience. By recognizing alterations in such rhythms, cities can not only anticipate but also adapt to pressing environmental challenges, such as water scarcity or biodiversity depletion.

Barbara Adam and Ecological Rhythms

In *Timescapes of Modernity*, Barbara Adam challenges us to reconsider the prevailing knowledge of linear, clock-based time, particularly in how we understand and engage with socio-environmental phenomena (Adam, 2005). She compellingly illustrates how the consideration of industrial clock time not only oversimplifies but also disrupts the rich interplay between the 'irreversible temporalities of life and the multiple rhythmicities of nature' (Adam, 2005, 95). This oversight ultimately neglects the complex temporalities that bridge nature and social life.

For Adam, this conventional approach is 'both part of the problem and applied as a solution' in industrial activities (Adam, 2005, 117). It construes time merely as a 'quantifiable resource' based on 'repetition without change' (Adam, 1998, 101), which effectively alienates us from the rich, multi-dimensional temporality inherent in natural processes and the varying contexts of living systems. This view fosters a reliance on standardized, quantified and decontextualized approaches in socio-technical systems that deeply undermine urban sustainability and environmental health.

In an effort to overcome these limitations, Adam's concept of timescapes offers a compelling lens through which we can explore intricate 'approaches to time and the multiple intersections of the times of culture and the socio-physical environment' (Adam, 2005, 122). Timescapes not only capture the 'rhythmicities, timings and tempos' that permeate both natural and cultural processes, but also illuminate the profound relationships and interactions that shape socio-environmental and socio-technical processes.

The notion of timescapes shifts our perception of time from that of a mere abstract, commodified resource to a rich, lived experience. It serves as a powerful tool, facilitating a 'practical engagement with the less tangible phenomena and processes of the global environment'. By delving into the temporal 'relationships, interdependencies and embeddedness' that drive social change (Adam, 2008, 7), Adam encourages us to recognize the diversity of the rhythms influencing socio-environmental phenomena and to determine which temporalities are prioritized and favoured. Acknowledging these rhythms is crucial for cultivating sustainable practices that honour the delicate timings of the natural world.

Lefebvre's rhythmanalysis, Gandy's space-time geography and Adam's timescapes invite us to dive into the interplay of space and time dynamics that shape urban socio-environmental phenomena. Revisiting these conceptualizations is crucial to see social, ecological and infrastructural rhythms as interconnected – a perspective that can inspire new dialogues between arts and sciences. Lefebvre's work not only unveils the various linear and cyclical rhythms in the urban environment, but also enriches our understanding of the diverse experiences arising from their interactions. Gandy expands this perspective by examining how urban systems can inadvertently exacerbate the disconnect between social, ecological and infrastructural rhythms. He urges us to navigate and question the intricate complexities of urban environments and the socio-ecological dynamics that dwell within. Meanwhile, Adam broadens our perspective by shedding light on the diverse temporalities that are often overlooked and unrecognized within socio-environmental processes.

Furthermore, with their conceptualizations, the authors touch upon different temporal scales of rhythms (Figure 2). While we can suggest that the social rhythms Lefebvre deals with are mainly visible in a daily and weekly setting, the infrastructural rhythms that Gandy addresses are most influential on monthly and seasonal scales. Finally, the ecological rhythms that Adam discusses in her work occur at broader temporal scales, such as yearly. For this reason, they open the doors for researchers to engage with diverse temporalities and understand their relationships, which creates the opportunity to bring arts and sciences closer.

Methodological Approaches to Rhythmanalysis in Social, Ecological and Infrastructural Processes

Delving into the intricate rhythms of biodiversity within urban ecosystems through the perspectives of Lefebvre, Gandy and Adam invites a nuanced exploration of diverse methodologies. Both artistic and scientific approaches can contribute to revealing the complex, interwoven dynamics of social, ecological and infrastructural processes. What can such methodologies look like? In this section, I will propose four methodological approaches for this: phenomenological and participatory approaches, documentary filmmaking and data visualization.

Phenomenological Approaches

Lefebvre posits a phenomenological approach to capturing the multifaceted rhythms across social, ecological and infrastructural processes, where observational techniques are imperative (Crang, 2001; Edensor, 2010). Observing rhythms is as much a systematic documentation of daily life patterns through scientific methods like ethnographic fieldwork as it is a way to attune to the sensory and experiential aspects of time and space – a process that lends itself to artistic practices.

For social rhythms, observations may focus on people's activities in outdoor spaces like parks, plazas and streets over a daily or weekly timeframe, especially at the neighbourhood or street level (Low, 1996). For ecological rhythms, we can centre observations around seasonality, documenting 'the regular and periodic changes of a condition on an annual scale' (White & Hastings, 2020). Likewise, for infrastructural rhythms, observations may focus on urban systems like roads in relation to environmental variables like air pollution across daily, weekly, or yearly temporal scales (Otu et al., 2024). Social, ecological and infrastructural rhythms may be visualized in the form of maps, timelines and charts. They can also be represented through photography, heat maps, network graphs, icons and diagrams (Simpson, 2012).

In observing rhythms, walking prevails as an effective research methodology in arts and sciences. As 'distinct senses of place do depend on the sensory experiencing of built environments', walking can enable researchers to experience the city 'through multiple sensory modalities, not just the visual', such as integrating touch, sound and smell (Degen & Rose, 2012). This enables the integration of not only scientific approaches but also artistic ones, which can alter the exploration of 'embodied inhabitation of urban spaces', connecting sensory experiences to shifting urban rhythms (Degen & Rose, 2012). Walking can also reveal 'how urban infrastructure lends identity to the urban landscape', documenting both the spatial orientation of infrastructures as well as their trajectories through time (Castán Broto et al., 2021). Gandy brings another perspective to these methods by introducing 'listening' and 'staying put' as additional dimensions to observing social, ecological and infrastructural rhythms (Gandy, 2024).

Technology-based approaches to these techniques may include the usage of sensors to document movement, climate-related patterns, and infrastructural usage patterns (Gabrys, 2019). Researchers have also used sound documentation to reveal the intersection of social, ecological and infrastructural rhythms, leading to the creation of soundscapes (Arkette, 2004). Both these methods are widely used in artistic and scientific research, since they enable rigorous documentation as well as opening an experiential and sensorial perspective to data.

Experimenting with different methods to capture social, ecological and infrastructural rhythms through observations allows artistic and scientific researchers to improve their 'spatial sensitivities', drawing attention to the 'lived experiences, behavioural changes, social values and practices' in relation to sustainability and climate change (Affolderbach & O'Neill, 2024). In this sense, observational techniques serve not only as a scientific tool but as a creative lens to capture how rhythms unfold across various spatio-temporal scales.

Participatory Approaches
Building on Adam's work, participatory and community-based methods offer another avenue to blend artistic and scientific research practices. Research on social and ecological resilience increasingly highlights the importance of participatory and transdisciplinary approaches, empowering citizens to co-create 'their adaptive capacity within a local urban area' (Champlin et al., 2023). Adam emphasizes the importance of lived, experiential aspects of socio-environmental phenomena, which participatory methods can capture (Facer et al., 2022). These methods not only complement observational techniques but also highlight the invaluable perspective of residents and local stakeholders, facilitating the interpretation of the unique rhythms intrinsic to their lived experiences (Calheiros et al., 2000). The participatory approaches not only capture local phenomena through empirical lenses, but also unveils the localized relationships between the informants and their surroundings.

In examining social rhythms, participatory research techniques such as interviews, workshops and focus groups can serve as vital platforms for residents to share their insights into their daily routines and community practices and the dynamics within their outdoor spaces (Springer & Skolarus, 2019). Artistic methods can strengthen these techniques by providing different ways for the participants to express themselves and narrate their experiences. When exploring infrastructural rhythms, creative methodologies like participatory mapping techniques or collective journaling projects can elucidate everyday patterns that animate specific locations. Additionally, participatory surveys and workshops invite residents to share how they interact with infrastructures, such as water supply or waste collection.

Such methods offer researchers and artists alike a richer understanding of how these social rhythms intertwine with local infrastructures and environmental contexts (Fagerholm et al., 2021). Furthermore, through such methods, we can better understand how urban residents respond to climate change or vice versa and what habits and behavioural patterns can be seen as actions that are beneficial from the sustainability perspective (Kurz et al., 2015).

Documenting ecological rhythms through community engagement may involve using creative tools, such as photography and diaries, to track seasonal transformations in their environments and articulate personal reflections on how these natural cycles interlace with their daily lives (Spoon, 2014). To deepen this engagement, workshops and focus groups can bring together residents and researchers, creating a space for discussion about local experiences with climate change, such as shifts in precipitation or temperature variations (Picketts et al., 2012). These collaborative methods, which are rooted in both arts and sciences, often illuminate hidden rhythms, unveiling culturally specific socio-ecological practices or adaptive strategies cultivated in response to local climate-related changes (Berkes & Turner, 2006).

Documentary Filmmaking

Gandy's analysis of the effects of infrastructures on socio-ecological processes highlights the critical role that artistic practices play in illuminating the intricate ways they interact across various scales (Gandy, 2022). Infrastructural rhythms provide an opportunity to weave together social and ecological narratives. Gandy's analysis of infrastructures such as water and energy supply and waste collection allow us to see these systems not just as technical constructs but as integral elements within complex social and ecological networks (Gandy, 2022). To engage meaningfully with the 'polyrhythmic account' of infrastructures, we need to examine specific rhythms and 'map its associations with other sites of rhythmic production', viewing infrastructure as a 'polyrhythmic assemblage' (Oppermann et al., 2020). This approach encourages exploration of the diverse rhythms of bodies and materialities, using artistic accounts alongside scientific methods.

In this light, Gandy highlights documentary filmmaking as a research method that enables a unique 'narrative experimentation as well as different modes of research dissemination' (Gandy, 2021). While documentary film enables 'a wider reflection on the relationship between cultural artefacts and materiality, . . . the affinity between the geohumanities and visual methods opens up possibilities for greater narrative experimentation as well as different modes of research dissemination' (Gandy, 2021).

This dual role of documentary film, as a powerful testament to lived experiences and a medium for rich intersubjective communication, makes it uniquely poised for exploring rhythms (Gandy, 2021). For example, capturing the interplay of ecological, social and infrastructural rhythms in a city through film might reveal how residents navigate the challenges posed by infrastructural failures or ecological shifts in their environment, transforming ordinary disruptions into compelling narratives of resilience. More than just a storytelling tool, the documentary film technique resonates deeply with audiences, evolving into a 'cultural artefact' that both reflects and influences broader discourses surrounding urban rhythms and sustainability (Gandy, 2021). Thus, documentary filmmaking emerges as a crucial artistic and scientific method in rhythmanalysis, inviting us to engage more profoundly with the rhythms of urban life and their wider significance.

Data Visualisation
Visualizing rhythms through using different forms of datasets offers a unique method for bridging the gap between arts and sciences, providing a platform for integrating both quantitative and qualitative approaches to understand urban socio-ecological processes. These datasets may include information on social, ecological and infrastructural rhythms collected both top-down and bottom-up. While urban spatial data on functions and streets gathered by municipalities can give hints about the social rhythms in a city, data on climate patterns and environmental qualities can enable knowledge of ecological rhythms. Finally, datasets on energy consumption, transportation flows, and water usage make it possible to understand infrastructural rhythms. Although these datasets are usually analyzed through scientific methodologies, engaging with artistic methods can enable the visualization of these data in creative ways.

When we delve into the complexities of socio-ecological processes, urban biodiversity and sustainability, recognizing the multifaceted scales at play is crucial (Faeth et al., 2012). In urban areas, the representation of these polyrhythmic aspects through different forms of data not only expands the experiential accounts behind socio-environmental processes, but also reveals pressing social and biodiversity needs. For instance, visualizing air pollution over time compared to human activity and infrastructural patterns may highlight disparities in access to healthy living environments (Adam, 2008).

Such visualizations also reflect how species and organisms respond to seasonal temperature changes and environmental conditions – processes whose spatial distribution and dynamics often demand long-term observation (White & Hastings, 2020) or are difficult to capture through participatory approaches. Integrating quantitative methods to capture broad spatial phenomena, like regional climate patterns, with qualitative approaches to uncover local-scale dynamics, such as community resource preferences, provides a holistic lens for addressing urban challenges (Albuquerque et al., 2020).

Focusing on the spatial context is just one aspect of data visualization of rhythms. By delving into the notions of time and temporality, we can unveil a richer understanding of the intricate dynamics at play in urban socio-environmental processes. Adam considers that time 'through this analytic promises to elucidate the ways in which political, social and economic conditions shape and exert authority over the everyday urban' (Besedovsky et al., 2019). This perspective invites scientists and artists to notice diverse temporalities unfolding in the urban context. These temporalities are not only anchored in a 'socio-historical past but also project into a socio-environmental future' (Adam, 2008). Temporal qualities surpass basic rhythmic measures like frequency, tempo and duration (Blue, 2019); they also open a window to the unknown future through 'social, familial, generational, economic, political and socio-technical' timelines (Adam, 2008). Exploring the alignment and misalignment between past, present and future, as well as daily, weekly and seasonal rhythms, is crucial to understanding how socio-environmental processes emerge from relations and interactions of various rhythms.

Bringing It Together

These approaches for capturing, analyzing and visualizing rhythms pave the way for a rich, multi-layered analysis, allowing us to delve into the intricate rhythms of socio-environmental dynamics. While scientific methods offer vital insights into the notions that shape social, ecological and infrastructural patterns, artistic methodologies ground these large trends in the lived experiences of individuals. By intertwining these approaches, we can construct a novel framework that examines the interplay of social, ecological and infrastructural rhythms in a dynamic, interconnected way.

The conceptual developments surrounding Lefebvre's rhythmanalysis, Adam's timescapes and Gandy's space-time geography illuminate how social, ecological and infrastructural rhythms shape urban environments and how they harmonize and clash in the urban context. Lefebvre's work serves as a vital lens to explore the dynamic qualities of urban life (Lefebvre, 2013), positioning the city as composed of cyclical and linear rhythms. Adam's notion of timescapes invites researchers to probe the complex layers of time that impact ecological, social and infrastructural rhythms (Adam, 2005). Meanwhile, Gandy sheds light on urban infrastructure and socio-economic processes, highlighting the critical spatio-temporal characteristics that give rise to these rhythms (Gandy, 2022).

In methodological terms, integrating these perspectives requires deploying scientific and artistic research methods to elucidate the interrelation of social, ecological and infrastructural rhythms. Phenomenological approaches, echoing Lefebvre's rhythmanalysis, enable researchers to discern recurring dynamics and their variations in daily life. Participatory approaches resonate with Adam's notion of timescapes, which invite urban residents to share their local knowledge and experiences related to the rhythms and temporalities that shape socio-environmental phenomena, offering a grassroots perspective to rhythmanalysis. Furthermore, Gandy proposes documentary filmmaking as a method that enriches the narratives of urban infrastructures and the various rhythms they entail. Finally, visualization of data from top-down and bottom-up sources appears to be an effective way to explore social, ecological and infrastructural patterns on different scales, which may be difficult to capture through qualitative forms of analysis.

While these approaches enable different perspectives on social, ecological and infrastructural rhythms, they also incorporate knowledge on different spatio-temporal scales (Figure 3). We can see the phenomenological approach through methods like observations focusing on a daily setting and on a specific spatial area. Similarly, the participatory approaches addresses a given community or a group of participants that will give evidence of a specific spatio-temporal context. In comparison to this, the documentary filmmaking approach will enable the registration of a larger spatio-temporal setting, integrating large geographies and times. Finally, data visualization may cover a broader space-time scale, enabling reflection on historical past and future predictions. This way, the four approaches together acknowledge the 'inherent temporal dynamics of urban systems', 'short-term planning responses', as well as 'long-term solutions' (Krishnan et al., 2024).

Rhythmanalysis is not only essential for bringing together diverse forms of knowledge, but also to reveal alignments and misalignments of spatio-temporal dynamics (Nevejan & Sefkatli, 2020). By correlating social, ecological and infrastructural rhythms, scientists and artists alike can identify points of alignment and misalignment, harmony and discord, eurhythmia and arrhythmia (Lefebvre, 2004) – insights that are essential for navigating urban ecological and infrastructural systems. Considering Lefebvre's work, exploring alignments and misalignments would mean looking at how linear and cyclical rhythms clash. In this perspective, Gandy encourages us to question how infrastructural rhythms intersect with social and ecological patterns (Gandy, 2022). On the other hand, Adam's notion of timescapes highlights the importance of recognizing and comparing rhythms on different spatio-temporal scales such as daily, seasonal and yearly, and across spatial extents, from neighbourhood streets to city districts and entire regions (Adam, 2005). By bringing these rhythms together, we weave a compelling narrative that juxtaposes these diverse scales, fostering a deeper appreciation and understanding of their interconnectedness.

The potential of merging the social, ecological and infrastructural rhythms into a cohesive analysis through these techniques inspires interdisciplinary and transdisciplinary collaborations that unite the realms of art and science. Rhythmanalysis is inherently based on 'an interdisciplinary approach', bringing together 'very diverse practices and very different types of knowledge' (Conlon, 2010). It equips researchers and practitioners with innovative tools and vocabulary that foster new understandings and methodological practices for interdisciplinary projects (Matos Wunderlich, 2024).

Ultimately, by articulating these complex relationships through creative output, we can identify where interventions are necessary. These creative processes enhance scientific analyses by allowing both experts and non-experts to engage in meaningful dialogues around urban sustainability, planning and development (Fokdal, 2021). In this context, Gandy asserts that through rhythmanalysis, we can create 'a critical entry point' for fostering a 'sustained intellectual dialogue' across disciplines (Gandy, 2023). This synthesis is crucial for developing urban infrastructural and planning interventions that not only cater to the social needs of communities but also champion the ideals of urban biodiversity and sustainability.

The Importance of Engaging Arts and Sciences through Rhythmanalysis for Sustainability and Biodiversity

This paper explored the rhythms within the social, ecological and infrastructural systems, shedding light on how the integration of scientific and artistic research methods can deepen our understanding of their complex interconnections in socio-environmental processes. While bringing the interplay between these dynamics to the surface can be challenging, it simultaneously opens doors to creative opportunities to bridge methods from arts and sciences in spatial research and design (Picketts et al., 2012). In this context, rhythmanalysis emerges as a 'practical, interdisciplinary method of analyzing urban temporal environments', producing invaluable knowledge that fuels the 'regenerative metabolism of cities' (Matos Wunderlich, 2024).

Looking forward to the future of spatial research and design in the context of climate adaptation and biodiversity in cities, we find ourselves at a crossroads of art and sciences in developing novel tools and approaches. By integrating artistic and scientific methods, rhythmanalysis stands out as a pivotal interdisciplinary approach that elucidates the complex temporalities inherent in urban environments. It compels both scientists and artists to rethink urban life through a nuanced and inclusive framework, which not only appreciates the aesthetic and experiential dimensions of rhythms, but also critically examines their implications on social, ecological and infrastructural processes. In this sense, rhythmanalysis serves as a 'boundary object', facilitating collaboration across different realms (Nevejan & Sefkatli, 2020) around design interventions that are attuned to the dynamics of human and ecological life.

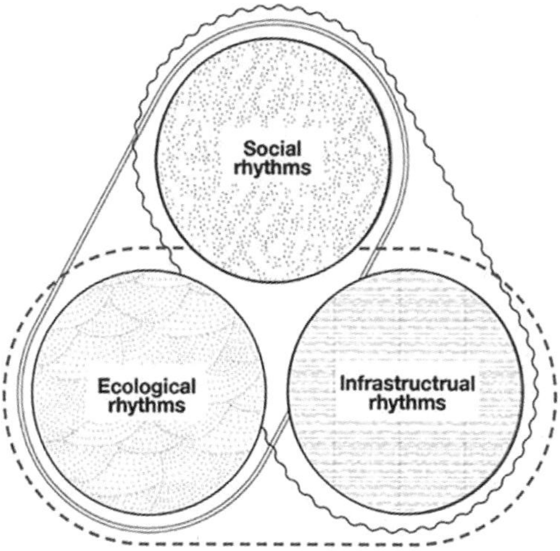

Figure 1: The interplay of the social, ecological and infrastructural rhythms. Pınar Şefkatli, 2024.

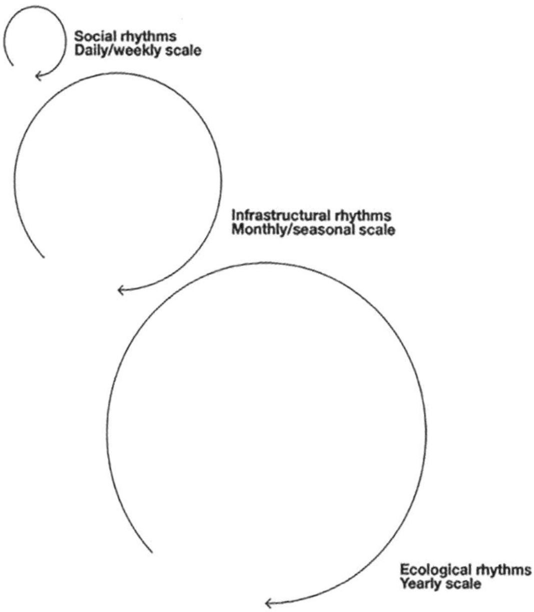

Figure 2: The different timescales in which social, ecological and infrastructural rhythms are manifested. Pınar Şefkatli, 2024.

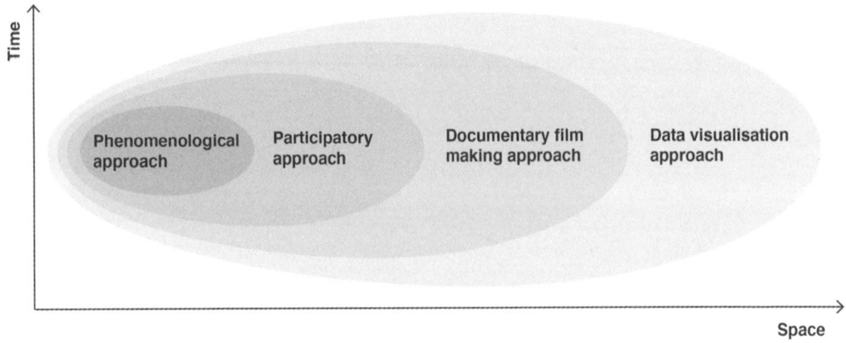

Figure 3: The various research approaches and their spatio-temporal impact. Pınar Şefkatli, 2024.

References

Adam, B. (2005). *Timescapes of Modernity: The Environment and Invisible Hazards* (London: Routledge).

Adam, B. (2008). 'The Timescapes Challenge: Engagement with the Invisible Temporal', in B. Adam et al. (eds.), *Researching Lives Through Time: Time, Generation and Life Stories*, 7-12 (Leeds: University of Leeds).

Affolderbach, J., & O'Neill, K. (2024). 'Everyday Sustainability Transitions through Using Green Buildings: Spatial Perspectives on Materialities, Discourses, and Lived Sustainabilities', *European Urban and Regional Studies*, *31*(2), 168-183. https://doi.org/10.1177/09697764231216407

Albuquerque, U. P., Ludwig, D., Feitosa, I. S., De Moura, J. M. B., De Medeiros, P. M., Gonçalves, P. H. S., Da Silva, R. H., Da Silva, T. C., Gonçalves-Souza, T., & Ferreira Júnior, W. S. (2020). 'Addressing Social-Ecological Systems across Temporal and Spatial Scales: A Conceptual Synthesis for Ethnobiology', *Human Ecology*, *48*(5), 557-571. https://doi.org/10.1007/s10745-020-00189-7

Arkette, S. (2004). 'Sounds Like City', *Theory, Culture & Society*, *21*(1), 159-168. https://doi.org/10.1177/0263276404040486

Berkes, F., & Turner, N. J. (2006). 'Knowledge, Learning and the Evolution of Conservation Practice for Social-Ecological System Resilience', *Human Ecology*, *34*(4), 479-494. https://doi.org/10.1007/s10745-006-9008-2

Besedovsky, N., Grafe, F.-J., Hilbrandt, H., & Langguth, H. (2019). 'Time as Infrastructure: For an Analysis of Contemporary Urbanization', *City*, *23*(4-5), 580-588. https://doi.org/10.1080/13604813.2019.1689726

Blue, S. (2019). 'Institutional Rhythms: Combining Practice Theory and Rhythmanalysis to Conceptualise Processes of Institutionalisation', *Time & Society*, *28*(3), 922-950. https://doi.org/10.1177/0961463X17702165

Brighenti, A. M., & Kärrholm, M. (2018). 'Beyond Rhythmanalysis: Towards a Territoriology of Rhythms and Melodies in Everyday Spatial Activities', *City, Territory and Architecture*, *5*(1), 4. https://doi.org/10.1186/s40410-018-0080-x

Butler, C. (2023). 'Autogestion and Ecological Politics in the Work of Henri Lefebvre', *Dialogues in Human Geography*, *13*(3), 457-460. https://doi.org/10.1177/20438206231157914

Calheiros, D. F., Seidl, A. F., & Ferreira, C. J. A. (2000). 'Participatory Research Methods in Environmental Science: Local and Scientific Knowledge of a Limnological Phenomenon in the Pantanal Wetland of Brazil', *Journal of Applied Ecology*, *37*(4), 684-696. https://doi.org/10.1046/j.1365-2664.2000.00524.x

Castán Broto, V., Sudhira, H. S., & Unnikrishnan, H. (2021). 'Walk the Pipeline: Urban Infrastructure Landscapes in Bengaluru's Long Twentieth Century', *International Journal of Urban and Regional Research*, *45*(4), 696-715. https://doi.org/10.1111/1468-2427.12985

Champlin, C., Sirenko, M., & Comes, T. (2023). 'Measuring Social Resilience in Cities: An Exploratory Spatio-Temporal Analysis of Activity Routines in Urban Spaces during Covid-19', *Cities*, *135*, 104220. https://doi.org/10.1016/j.cities.2023.104220

Conlon, D. (2010). 'Fascinatin' Rhythm(s): Polyrhythmia and the Syncopated Echoes of the Everyday', in T. Edensor (ed.), *Geographies of Rhythm: Nature, Place, Mobilities and Bodies*, 71-81 (Farnham: Ashgate).

Crang, M. (2001). 'Rhythms of the City: Temporalised Space and Motion', in J. May & N. Thrift (eds.), *Timespace: Geographies of Temporality*, 187-207 (London: Routledge).

Degen, M. M., & Rose, G. (2012). 'The Sensory Experiencing of Urban Design: The Role of Walking and Perceptual Memory', *Urban Studies*, *49*(15), 3271-3287. https://doi.org/10.1177/0042098012440463

Edensor, T. (2010). *Geographies of Rhythm: Nature, Place, Mobilities and Bodies* (Farnham: Ashgate).

Facer, K., Siebers, J. I., & Smith, B. (eds., 2022). *Working with Time in Qualitative Research: Case Studies, Theory and Practice* (London: Routledge).

Faeth, S. H., Saari, S., & Bang, C. (2012). 'Urban Biodiversity: Patterns, Processes and Implications for Conservation', in: *Encyclopedia of Life Sciences* (Hoboken, NJ: Wiley-Blackwell). https://doi.org/10.1002/9780470015902.a0023572

Fagerholm, N., Raymond, C. M., Olafsson, A. S., Brown, G., Rinne, T., Hasanzadeh, K., Broberg, A., & Kyttä, M. (2021). 'A Methodological Framework for Analysis of Participatory Mapping Data in Research, Planning, and Management', *International Journal of Geographical Information Science*, *35*(9), 1848-1875. https://doi.org/10.1080/13658816.2020.1869747

Fokdal, J., Bina, O., Chiles, P., Ojamäe, L., & Paadam, K. (eds., 2021). *Enabling the City: Interdisciplinary and Transdisciplinary Encounters in Research and Practice* (London: Taylor & Francis).

Foster, J. B., Napoletano, B. M., Clark, B., & Urquijo, P. S. (2020). 'Henri Lefebvre's Marxian Ecological Critique: Recovering a Foundational Contribution to Environmental Sociology', *Environmental Sociology*, *6*(1), 31-41. https://doi.org/10.1080/23251042.2019.1670892

Gabrys, J. (2019). 'Sensors an Sensing Practices: Reworking Experience across Entities, Environments, and Technologies', *Science, Technology, & Human Values*, *44*(5), 723-736. https://doi.org/10.1177/0162243919860211

Gandy, M. (2005). 'Cyborg Urbanization: Complexity and Monstrosity in the Contemporary City', *International Journal of Urban and Regional Research*, *29*(1), 26-49.

Gandy, M. (2014). *The Fabric of Space: Water, Modernity, and the Urban Imagination* (Cambridge, MA: The MIT Press).

Gandy, M. (2021). 'Film as Method in the Geohumanities', *GeoHumanities*, *7*(2), 605-624. https://doi.org/10.1080/2373566X.2021.1898287

Gandy, M. (ed., 2022). *Natura Urbana: Ecological Constellations in Urban Space* (Cambridge, MA: The MIT Press).

Gandy, M. (2023). 'Chennai Flyways: Birds, Biodiversity, and Ecological Decay', *Environment and Planning E: Nature and Space*, *6*(4), 2678-2699. https://doi.org/10.1177/25148486221142491

Gandy, M. (2024). 'Attentive Observation: Walking, Listening, Staying Put', *Annals of the American Association of Geographers*, *114*(7), 1386-1404. https://doi.org/10.1080/24694452.2024.2353841

Heynen, N., Kaika, M., & Swyngedouw, E. (2006). 'Urban Political Ecology. Politicizing the Production of Urban Natures', in N. Heynen, M. Kaika, & E. Swyngedouw (eds.), *In the Nature of Cities: Urban Political Ecology and the Politics of Urban Metabolism*, pp. 1-20 (London: Routledge).

Jones, O. (2010). "The Breath of the Moon': The Rhythmic and Affective Time-Spaces of UK Tides', in T. Edensor (ed.), *Geographies of Rhythm: Nature, Place, Mobilities and Bodies*, 189-203 (Farnham: Ashgate).

Krishnan, S., Aydin, N. Y., & Comes, T. (2024). 'TIMEWISE: Temporal Dynamics for Urban Resilience – Theoretical Insights and Empirical Reflections from Amsterdam and Mumbai', *Npj Urban Sustainability*, *4*(1), 4. https://doi.org/10.1038/s42949-024-00140-5

Kurz, T., Gardner, B., Verplanken, B., & Abraham, C. (2015). 'Habitual Behaviors or Patterns of Practice? Explaining and Changing Repetitive Climate-Relevant Actions', *WIREs Climate Change*, *6*(1), 113-128. https://doi.org/10.1002/wcc.327

Lefebvre, H. (1996). *Writings on Cities* (trans. E. Kofman & E. Lebas). (Hoboken, NJ: Wiley-Blackwell).

Lefebvre, H. (2004). *Rhythmanalysis. Space, Time and Everyday Life* (trans. S. Elden & G. Moore). (London: Continuum).

Lefebvre, H.; ed. D. Nicholson-Smith (2013). *The Production of Space* (Hoboken, NJ: Wiley-Blackwell).

Lefebvre, H. (2014). *Critique of Everyday Life: The One-Volume Edition* (London: Verso).

Low, S. M. (1996). 'Spatializing Culture: The Social Production and Social Construction of Public Space in Costa Rica', *American Ethnologist*, *23*(4), 861-879. https://doi.org/10.1525/ae.1996.23.4.02a00100

Matos Wunderlich, F. (2013). 'Place-Temporality and Urban Place-Rhythms in Urban Analysis and Design: An Aesthetic Akin to Music', *Journal of Urban Design*, *18*(3), 383-408. https://doi.org/10.1080/13574809.2013.772882

Matos Wunderlich, F. (2024). *Temporal urban design: Temporality, rhythm and place* (London: Routledge).

Nash, L. (2020). 'Performing Place: A Rhythmanalysis of the City of London', *Organization Studies*, *41*(3), 301-321. https://doi.org/10.1177/0170840618789161

Nevejan, C., & Sefkatli, P. (2020). 'City Rhythms: An Approach to Urban Rhythm Analysis', in P. Crespi & S. Manghani (eds.), *Rhythm and Critique: Technics, Modalities, Practices*, 218-241 (Edinburgh: Edinburgh University Press).

Oppermann, E., Walker, G., & Brearley, M. (2020). 'Assembling a Thermal Rhythmanalysis: Energetic Flows, Heat Stress and Polyrhythmic Interactions in the Context of Climate Change', *Geoforum*, *108*, 275-285. https://doi.org/10.1016/j.geoforum.2019.09.012

Otu, E., Ashworth, K., & Tsekleves, E. (2024). 'Rhythm of Exposure in Town Centres: A Case Study of Lancaster City Centre', *Environments*, *11*(7), 132. https://doi.org/10.3390/environments11070132

Picketts, I. M., Werner, A. T., Murdock, T. Q., Curry, J., Déry, S. J., & Dyer, D. (2012). 'Planning for Climate Change Adaptation: Lessons Learned from a Community-Based Workshop', *Environmental Science & Policy*, *17*, 82-93. https://doi.org/10.1016/j.envsci.2011.12.011

Popke, J. (2016). 'Researching the Hybrid Geographies of Climate Change: Reflections from the Field', *Area*, *48*(1), 2-6. https://doi.org/10.1111/area.12220

Sefkatli, P. (2024). 'Rhythm Matters! How Rhythm Analysis Bridges Architecture and Sociology', *archiDOCT*, *12*(1). https://openresearch.amsterdam/image/2024/7/23/118893_rhythm_matters_how_rhythm_analysis_bridges_architecture_and_sociology.pdf

Simpson, P. (2012). 'Apprehending Everyday Rhythms: Rhythmanalysis, Time-Lapse Photography, and the Space-Times of Street Performance', *Cultural Geographies*, *19*(4), 423-445. https://doi.org/10.1177/1474474012443201

Spoon, J. (2014). 'Quantitative, Qualitative, and Collaborative Methods: Approaching Indigenous Ecological Knowledge Heterogeneity', *Ecology and Society*, *19*(3), art33. https://doi.org/10.5751/ES-06549-190333

Springer, M. V., & Skolarus, L. E. (2019). 'Community-Based Participatory Research: Partnering with Communities', *Stroke*, *50*(3). https://doi.org/10.1161/STROKEAHA.118.024241

White, E. R., & Hastings, A. (2020). 'Seasonality in Ecology: Progress and Prospects in Theory', *Ecological Complexity*, *44*, 100867. https://doi.org/10.1016/j.ecocom.2020.100867

Dialogue with Paul Shepheard

Paul Shepheard and Sophia Arbara

Reconstructing Narratives

Sophia Arbara: One of the goals of the Reporting the Delta project is to unveil how – in the past – specific technical solutions have been able to affect current social, spatial, and ecological conditions. The project does so by using less academic means, such as historical documentaries, as a pretext to bridge research between arts and sciences and move scientific knowledge on technological design and environmental changes closer to citizens. In your own books, you also use a narrative style with a strong personal tone, for example to address the idea of cultivated wilderness across different conditions, uncommon in academic writings. Why did you choose this approach?

Paul Shepheard: The universities are pursuing the truth. But because they use words to do it they are continually faced with this problem: writing is an approximation, not a certainty. When you write, you are always translating. No matter what protocols and guarantees you construct, the elasticity of words will remain, and it is there that I find my own voice. The complication is that what I write about is the certainties of the material world. Things exist in the world in relation to other things and forces as a condition of existence. What writers do is select from the complex infinity of things in the world and arrange them into contexts to establish meaning; a context is a set of relationships with meaning and narrative is a description of a context.

Whatever protocols and curriculums are erected, there will always be an ambiguity in words that compromises them and turns them into imitations of the truth. When I was working in Antwerp, I arranged to meet the head of school and his head of history to discuss whether they would publish my fourth book *How to Like Everything*. I took along the Cultivated Wilderness to show them and the history man flicked through it as if it was a catalogue and said, 'but there are no footnotes!' That's when I saw the gap between the academies and myself. The academy builds its search for truth upon a ground of corroboration. It asks that every contribution fits into the tower of knowledge under construction. In contrast, I liken my own thoughts to a pack of nomadic animals that live outside the stasis of the tower. The leaders of these fluid organizations cannot rely on precedent or law, since everything – everyone – is always shifting. They need practical cunning. As Deleuze and Guattari put it , 'the leader has to gamble everything on every move' (Deleuze & Guattari, 1987). Now what I'm thinking is that I am the leader of a pack of thoughts, the pack of likes and dislikes that populates my head. This pack is continuously shifting, things on the periphery becoming central, things in the middle flying to the edge; I must let the pack shift and self-organize, I must not burden it with structure but follow the flow, and gamble on a positive outcome. I must do the opposite of what is done in the academies.

You also suggest that historical reinterpretation can reshape our understanding of landscapes and places. For example, you describe how the stories about horses and landscapes described in the Iliad and the Odyssey may not make sense in the Aegean topography but rather on the islands of the North Sea and the Atlantic and that Troy was a Celtic city, with Homer originating from Middelburg in the Dutch Delta. How important do you think it is to revisit known stories through different lenses, and what does this add to our perception of places?

The Garden of Eden story of the fall and Paradise Lost has an invigorating alternative in the Coptic heresy, which says that in the beginning, the created world was a spiritual phenomenon, but Satan fashioned it into the material world for Adam and Eve to live, where they would have no knowledge of good and evil but would experience pleasure and suffering. God sent a serpent to explain the danger they were in, which made Satan so angry that he smashed up paradise, and that broken world is where we live now, among the storms and the famine and disease. The glimpses of beauty we see in the world are the shards of the spiritual world, the pleroma, still stuck in the wreckage.

The prominent story now is one of a nature damaged by human actions that must be saved. It is a version of the Eden story, whose hold on our human imagination is still powerful. Even the theory of evolution, which explains material existence in another way, is treated as a creation story, with extinctions seen not as the engines of change but as part of the tragedy of the fall. Questions arise: why should we work, for example, to save parasitic behaviours like the cuckoo's? And the biggest question for me is, is life nature? Or does it work against nature, with its specifications, its separations and its greed for advancement?

As architects and designers we often do not limit ourselves in designing the physical space but produce and re-formulate past narratives such as the stories above but also speculate upon future scenarios. In *What is Architecture?*, you call yourself a speculator. What does this mean for you?

Speculation is about seeing, and trusting the evidence of your eyes. In *What is Architecture?* I use it to denote acts of the imagination. The story of Donatello pulling Roman statues intact out of the piles of rubble left behind by the Goths when they sacked the city a thousand years before, and thus kickstarting the renaissance, is probably, I can't remember but I guess, in Vasari. The point of calling it a speculation is that the impact of the story is more important than its provenance.

Within that context, how do you see the role of architects and designers in the environmental discourse?

Don't ask me this. Everyone has climbed on board this bandwagon. I have started to think of the 'environmental discourse' as a trap. It tends to privilege stasis over the dynamic flows of the world. What exists over what might exist. I guess the role of all of us is to maintain perception of the dynamics, The world is not a churchyard.

Utility Landscape in the Dutch Delta

Moving to the context of the Dutch delta and the performance of the landscape, around the 1950s the Delta has been characterized as a productive landscape. Is this similar to a utility landscape that you refer to for the Dutch setting? What is a utility landscape and is there beauty in this utility?

Utility means something that is useful. The word has philosophical pedigree in its use in the idea of utilitarianism. It is an early nineteenth century idea. It is a picture of democracy, where choices are made by election. Fifty one percent is the dividing line between action and rejection. Decisions made by

> at least fifty-one percent ensures the 'greatest happiness for the greatest number'. Dutch democracy is strong, and I think it derives this strength through the practical operations of the waterboards that define administrative boundaries. Hence, the 'utility' of the polder landscape. However, democracy is not a judgment, but a tool. It is neutral. So if the majority wants Hitler, that's what they get. This casts some light on your questions about recreation. Land, for all its distortions as a financial instrument, is fundamentally neutral.

On one hand the Dutch delta has become a utility landscape but also a heavily recreational landscape (leisure on the water, tourists, 'horsification'). Is this recreation also part of the utility landscape, in other words, is recreation also a utility? Do you think this shift has been recent in the Dutch delta from a landscape of production to a recreational one?

> This is what life does: It takes the conditions of the world as it finds them and exploits them for its own purposes. I suppose that the use of hard-won new land for recreation is a result of changing market forces – in which case, relax: nothing lasts forever.

Machines in the Landscape and Social Engineering

We all know that the Dutch landscape is a heavily constructed one with the Delta Works being an unprecedented work of human engineering, creating monumental machine landscapes. Do you think the creation of such machine landscapes came with a feeling of awe at the miracle of growth but also fear of our inability to estimate the power of machines to transform environments?

> This makes me think of other great engineering initiatives like the Great Wall of China, the construction of the Egyptian pyramids and the ring of twelfth-century gothic cathedrals around Paris. I'm not convinced that machine made is the most important quality of the Delta Project. Fear does play its part – I think you are right to draw attention to it. Frankenstein's monster is a parable of fear at the unleashing of scientific knowledge and the twentieth-century preoccupation with mechanization has been accompanied by fears of progress that survives in the anxieties about social media and artificial intelligence. On the other hand, there is maturity in the Room for the River program of unstitching the more inflexible elements of the project. That seems like working with nature to me – recognizing the awesome power of it. Gravity, hydrology, the sun's energy, the spinning planet: these are nature.

Endangered species is a different issue. We need to work with humans on that, not nature.

These infrastructures were developed at times where modernization embodied the conquest of nature, accompanied with positivity from the Dutch labour force that endured times of hard-working conditions in exchange of becoming settlers of a new land, seen as pioneers. To what extent has this narrative of national pride at the time of Delta Works played a key role in making the Dutch engineering know-how a product of export and selling it as a success story in other contexts?

I have no knowledge of this, except to say that I have been told that early polder settlers were called Turnip Heads by this nation of democrats – which is surprising.

The Cultivated Wilderness, the Landscape

In this last part of our conversation, we would like to ask you about the concept of cultivated wilderness. You describe in your work 'wilderness: the world before humans appeared in it', 'cultivation: everything we've done to it since' and 'landscape: the strategies that have governed what we've done'. You also mention that wilderness is everywhere yet at the same time, the whole world is cultivated and bears the signs of this cultivation. So, what is the cultivated wilderness and what is landscape in this context? Also, you describe the Dutch delta as an area with busily harvested farms, huge infrastructural delta works, trucks and containers moving goods all over the world. In such an intensively altered landscape by human interventions, where does the wilderness start and end?

I admit there is a weakness in my trope of cultivation. Binaries are useful, but they need tight control. You will guess that my position is slightly different than it was in 1996 when *The Cultivated Wilderness* was written. I want to assert that life is not nature, it's something else. Life differentiates, speciates, and builds. Nature is uniform, inevitable. It combines and decays. How I deal with this question about life not being nature should be the subject of another book. I hope I have time.

Do we need a post-wilderness concept of wilderness? And would that be the cultivated wilderness?

This is your best question. In the seventeenth century, travellers used to block out views of open country that they were passing through because of the frightening character of the wilderness, Thoreau was a pioneer in changing that view. I believe he used 'wildness' to describe his perception because of the pejorative that was contained in the old view of wilderness. Wildness is softer, more benign, more suited to the conversation about 'ecology'. What's not so benign is the rewilding programs that attempt to repopulate the land with wolves. Can you hear them howling? Does no one remember the stories that made us persecute wolves in the first place? Think on it.

References

Deleuze, G., & Guattari, F. (1987) [1980]. *A thousand plateaus: Capitalism and schizophrenia* (transl. B. Massumi). (Minneapolis: University of Minnesota Press).

Shepheard, P. (2013). *How to like everything: A utopia* (London: Zero Books).

Shepheard, P. (1997). *The cultivated wilderness: Or, what is landscape?* (Cambridge, MA: MIT Press).

Shepheard, P. (1994). *What is architecture?* (Cambridge, MA: MIT Press).

Thoreau, H. D. (1862). "Walking", *The Atlantic Monthly*. Available: https://www.theatlantic.com/magazine/archive/1862/06/walking/304674/ (accessed 10 April 2024).

De lage landen, George Sluizer, 1960

Rommy Albers

De lage landen (*Hold Back the Sea*) is an educational documentary commissioned by Shell and produced by George Sluizer, highlighting the history of the reclamation of the western part of the Netherlands. Sluizer begins by showing a landscape below sea level: living, working, recreation and agriculture mostly take place on land below sea level. With the help of dikes, canals, pumping stations and sluices, the Dutch have managed to protect their land from the water.

Sluizer then starts his historical overview in antiquity, showing that the western dune area of the Netherlands was made habitable by the construction of primitive dikes. The overview continues until around 1612, when the first reclamations took place using windmills. Especially in northern Holland – the areas between Leiden and Amsterdam and north of Amsterdam – large inland lakes such as the Beemster, Schermer and Purmer were reclaimed. In the nineteenth century, steam-powered pumping stations enabled the Haarlemmermeer, a large inland sea southeast of Haarlem, to be drained and rivers to be canalized. Thanks to the construction of the Nieuwe Waterweg, a canal between the harbour of Rotterdam and the North Sea, Rotterdam could become the largest port in the world. The Twentieth Century section of the film features the large-scale Zuiderzee Works with the construction of the Afsluitdijk and the impoldering of large parts of the IJsselmeer. The film ends with the construction of the Delta Works and a vision for the future, in which the whole of the Netherlands will be protected by concrete sea walls.

In *De lage landen*, Sluizer depicts the positive attitude that the land is engineerable, that the Dutch have succeeded in controlling the sea and water, and have brought the reclaimed land under cultivation. Even major disasters such as the Sint-Elisabethsvloed (1421) and the flood disaster of 1953 have not been able to turn the tide and have only led to the construction of new hydraulic works.

Grounding the Future in the Delta

Fransje Hooimeijer

Climate change demands a radical rethinking of delta management to ensure long-term safety and sustainability. The Dutch have a long tradition of balancing urban development and water management, using innovative approaches like polders, dikes and windmills to adapt to hydrological challenges. However, large-scale engineering interventions have weakened the connection between flood protection systems and regional water dynamics, emphasizing the need for integrated approaches (National Research Council, 2013). Drawing on the Redesigning Deltas study, this contribution examines how spatial design and civil engineering can address contemporary challenges by adopting a longue durée perspective. This approach aligns historical principles with modern strategies, emphasizing adaptive design, bottom-up governance and the importance of public engagement. Through case studies, it demonstrates how lessons from the past can inform innovative solutions for future delta management, balancing ecological, social and technical considerations.

Introduction

Climate change has heightened the urgency to radically rethink delta management practices, as current approaches cannot ensure long-term safety. Addressing this challenge requires building the knowledge to change existing methods and develop new approaches for managing deltas (Bento, 2023). The Dutch have a long-standing and internationally recognized tradition of effectively managing deltas, reflecting a strong connection between urban development and civil engineering. Their expertise in hydrological principles enabled land reclamation through the ingenious technology of polders (Burke, 1956). The dynamics of regional water systems – groundwater, rainwater and surface water – have been central to the urbanization of Dutch polders. Historically, cities in polders were hydrological constructions, with spatial layouts deeply intertwined with the division of land and water at regional and local scales (*ibid.*). However, advancements in site preparation and large-scale water pumping weakened this connection (Hooimeijer, 2014). Today, subsided polders and cities rely heavily on pumps and extensive dike systems for flood protection. To ensure sustainability, delta management must integrate flood protection systems with regional water dynamics, encapsulated in the concept of the *deltaic condition* (National Research Council, 2013).

This relationship between technical efficiency and urban design has evolved over time. Over the past millennium, societal needs and technological advancements have continually redefined the deltaic condition. As climate change exacerbates hydrological challenges, fostering a closer relationship between civil engineering and spatial design becomes imperative. Historical interactions between spatial design and wet landscapes offer principles that can guide future strategies, including site preparation, protection scales and land-use planning (Hooimeijer, 2014).

To explore new approaches to delta management, TU Delft collaborated with Deltares, Resilient Delta, Planbureau voor de Leefomgeving (PBL), Erasmus University and Wageningen University in the Redesigning Deltas study. The study investigated how spatial design and engineering can synergize to create a sustainable future for the Dutch delta. It emphasizes the role of design as a mediator and knowledge broker, helping to address climate uncertainties by envisioning alternative approaches (Hooimeijer, 2023). Design acts as a catalyst, aligning diverse ambitions, goals and financial considerations across disciplines to create spaces that are smarter, more sustainable and more affordable.

The study engaged fifteen partners from the urban planning, landscape architecture and water management sector, addressing five design challenges across the Dutch delta. Teams included Defacto, Vista and Arcadis working on Limburg's brook landscape;

Fabrications, Bosch Slabbers and Tauw in the Waal River region; Urbanisten, Lola and Royal Haskoning on Rotterdam's port-city interface; ZUS, Flux and Sweco in the polder between Rotterdam and Delft (Midden-Delfland); and Studio Hartzema, Feddes Olthof and Witteveen & Bos in Zeeland. These interdisciplinary teams collaborated in sessions with experts, developing strategies that integrated ecological, socio-economic and spatial challenges from a systems perspective.

The study's results, encompassing innovative designs and a manifesto, offer insights into the future of everyday life in the Dutch delta while drawing on historical practices. The manifesto emphasizes a bottom-up approach to delta management, mirroring pre-1814 practices, when local knowledge, smaller-scale technologies and path dependencies between land and water were pivotal (Van der Ham, 2002). These findings demonstrate how historical concepts can provide valuable frameworks for addressing contemporary challenges. By building on the synergy between spatial design and civil engineering, the *Redesigning Deltas* study proposes a way forward for delta management that harmonizes ecological, social and technical needs. Historical insights serve as a reminder that adaptive, localized solutions have long ensured resilience in dynamic environments.

This carnet will first give a summary of the historical principles of delta management to contextualize the result of the Redesigning Deltas Design study (RDD). The project by ZUS, Flux and Sweco in Midden-Delfland, which actively engaged with local stakeholders and extended the impact of the study beyond its initial scope, is highlighted to draw conclusions with insights into the future of everyday life in the Dutch delta.

Linking History of Delta Design to the Future

The historical relationship between civil engineering and spatial design in managing wet, soft soils and balancing land and water has provided essential spatial principles for site preparation, protection scales, and land-use planning (Hooimeijer, 2014). Over time, technical innovations have influenced the Dutch approach to water systems, evolving from an adaptive stance to one focused on minimizing risk. Van Dam (2016) describes an 'amphibious culture' until 1800, after which the dominant paradigm shifted to reducing risks to an absolute minimum. Learning from historical periods with limited technology offers valuable insights for developing more sustainable, nature-based solutions.

The history of Dutch water management can be divided into phases marked by technological advancements: Natural Water Management (ditches, until 1000), Defensive Water Management (dikes, 1000-1500), Anticipative Water Management (windmills,

1500-1800), Offensive Water Management (steam, 1800-1890), Manipulative Water Management (electricity, 1890-1990) and Adaptive Manipulative Water Management (return to nature, 1990 to present) (Hooimeijer, 2014). Each phase represents a shift in the balance between human ingenuity and the constraints of the natural environment.

During the phase of Natural Water Management, Dutch lowlands were marshlands, uninhabitable except when adaptations were made to the wet environment. Ditches were dug to control groundwater for agriculture, but there was no unified effort to protect against flooding. Settlements were established on higher ground, often near streams and fertile areas. A key typology of this period was the mound (*terp*), raised land used as refuge during floods, where communities maintained essential functions. These early adaptations laid the foundation for later advancements in water management (Burke, 1956).

The transition to Defensive Water Management began around 1000 with the introduction of dikes (Van der Ham, 2002). This innovation enabled the protection of larger areas, making them habitable while integrating water into settlement designs through harbours. Dike and dam cities flourished during the thirteenth and fourteenth centuries, with building sites along dikes raised using debris. These developments marked a significant evolution in balancing urbanization and flood protection.

By 1500, the use of windmills marked the shift to Anticipative Water Management. Windmills allowed for large-scale water movement, enabling the drainage of extensive areas and towns. This phase reflected a deeper understanding of hydrological systems and practical applications of that knowledge. Combined with sluices and dams, windmills transformed water management from a defensive to an anticipative stance. Settlements expanded onto previously unsuitable wet soils, driven by a forward-looking vision that anticipated future needs and incorporated technical expertise (Burke, 1956). The layout of early settlements – mound, river, coast, military stronghold (*burcht*), *geestgrond* (sandy ridges), dike and dam towns – formed the 'dry core' of what would become polder cities. These settlements prospered and expansion extended into surrounding wet soils initially used for agriculture. This process required collaborative vision and technical ingenuity, reflecting the hallmarks of the Dutch tradition in water management.

The Offensive Water Management era (1800-1890) was characterized by the use of steam power, enabling large-scale interventions such as canals, sea closures and artificial groundwater adjustments. These advances transformed Dutch cities into industrial hubs, concentrating populations around factories and harbours. Urbanization expanded rapidly, reshaping the landscape and pushing the boundaries of water management technology (Van der Woud, 1987).

By the late nineteenth century, rapid urbanization and new technologies defined the Manipulative Water Management era (1890-1990). The rise of engines, electricity and automobiles, coupled with industrialized construction processes, created a new spatial order. This era of *maakbaarheid* ('everything can be engineered') prioritized technological solutions, leading to the disconnection of water management from natural systems. The emphasis on industrial progress overshadowed the ecological balance, altering the relationship between urban planning and water management (Van der Woud, 1987).

During the Manipulative era, three distinct periods can be identified. In the Inter War Era, technology was still in tune with natural systems. Post-World War Two was an era where confidence in technology led to a complete separation between water and urban systems. From the 1970s onward, the post-war era was criticized for its technocratic and narrow-minded approaches to social structures. There was a strong desire to break free from these conventions and rediscover the 'real identity' of the city and countryside. Respect for nature became a central theme, leading to the emergence of landscape architects as key players in urban planning. They reintroduced water as a spatial element in the city, marking a shift towards more ecologically-aware planning.

This shift set the stage for the next phase: Adaptive Manipulation (1990 to today). Although the term 'adaptive manipulation' is contradictory, it reflects the ongoing debate about how to make the right spatial adjustments to adapt to climate change. There is still a lack of consensus among experts about what the 'right' adjustments are, which highlights the complexities of integrating technological advances with respect for natural systems in the face of an uncertain future.

The above-mentioned historical practices and principles can inform modern approaches to delta management; in particular, periods with limited technological resources can guide the development of more sustainable and nature-based solutions for the future. The main concepts derived from historical principles of water management in the Dutch delta revolve around several key ideas: (i) design in levels situating human occupation on higher ground to protect against flooding; (ii) protection at the regional and local scales, creating socially cohesive dike rings that offer protection tailored to specific communities; (iii) design with a vision for the future, anticipating future developments and incorporating them into current planning; (iv) working with the natural system, recognizing and respecting the natural boundaries and carrying capacities set by the environment, which dictate the limits of human occupation.

These principles were eventually set aside as new approaches emerged that were more aligned with the evolving attitudes, needs and expectations of society. The scaling up of water defence and management systems could, in hindsight, be seen as a form of 'disruptive innovation'. This shift fundamentally altered the deltaic system to such an extent that returning to the old ways of functioning is no longer possible. As a result, the future of the Dutch delta will likely involve a combination of hybrid solutions – integrating both old and new approaches – or potentially a complete transition that could drastically reshape, or even lead to the loss of, the traditional Dutch delta as we know it.

Overview and Impact of the Redesigning Deltas Design Study

The RDD study employed several methods to facilitate interdisciplinary collaboration and design (Hooimeijer, 2023). The group of participating firms played a key role in setting the assignment and were further supported through masterclasses and workshops, which provided the designers and engineers with specialized knowledge. The study emphasized the integration of disciplinary perspectives from urban design, landscape architecture and engineering, represented by the group members. This integration was achieved through a methodology known as a 'sandbox'. A sandbox is an interdisciplinary approach used to explore tasks broadly and inventory the necessary knowledge to arrive at a shared problem definition and approach. This method allows the collective practical experience of the group to be harnessed, building a preliminary collective understanding of the design challenges, shaping the design brief, and helping to further define initial propositions as a group. This was done on the first day of the sandbox meeting, and resulted in forming five interdisciplinary teams, each composed of an urban design, landscape architecture and engineering firm. The locations were then distributed among these teams. On the second day, the teams presented the strategies they had developed for their locations. Following these presentations, the disciplines reconvened within their disciplinary groups to draw conclusions at a broader scale, ensuring the five propositions were connected into a cohesive whole.

These propositions were further refined using the 'Casco Concept' – a design instrument developed in the 1990s to stimulate sustainability (Sijmons, 1991) – in a workshop facilitated by H+N+S Landscape Architects. H+N+S had been commissioned by the RDD study to revise its existing design tool, the Layer Approach – a planning tool to help making priorities on the basis of dynamic, prioritizing slow (the substratum layer) over fast, the occupation layer (De Hoog et al., 1998). H+N+S adapted the Layer

Approach and Casco Concept to make them applicable for future delta design (H+N+S, 2023). The workshop results led to a shared foundation for all five teams: the soil and water system, rather than land use, would form the basis of future developments. This approach ensured that all proposals were grounded in the same logic, culminating in a comprehensive plan for the entire southern Dutch delta.

This unified plan was supported by a manifesto advocating a 'new Dutch' design approach to flood risk management. The manifesto emerged from the participants' desire to create an independent, collective voice from the design discipline in discussions about developing a sustainable Delta system. The manifesto consists of six key propositions, three of which are about decision making:

- We are here to stay, we take responsibility and we can change
- We need to know more
- We dare to make painful decisions

The other three are on design:

- We utilize the delta paradox: regulation within the dynamics
- We design the delta bottom-up
- The design of the delta starts with the section

These propositions are represented in the work of the five teams and are considered as shared values needed to be able to design the delta of the future.

Midden-Delfland Project

The team Midden-Delfland with ZUS, Flux and Sweco investigated the relationship between city and polder. The water regimes in both land uses perform very differently. In the polder there are many different water levels, making it a complicated and inflexible system. The three partners envisage Midden-Delfland as a National Park in which water, food and raw materials are produced for the adjacent cities, which will remain within strict contours. The design is based on a simplified water system with fewer dikes, pumps and sluices, with opportunities for increasing biodiversity, supplying raw materials such as wood and food, and offering a solution to flooding, subsidence and the threat of salinization. As a result of population growth, Delft and Rotterdam are becoming more densified within the urban contours, increasing the need for nature, water and clean air. Midden-Delfland will become a green lung that contributes to the liveability of the highly densified urban environments along the edges. Those who live in the polder will be in very close relationship with the water system; they will no longer rely on large-scale public protection measures, as each house will be able to cope with all water conditions (Figure 1).

The project is reflective of the RDD manifesto proposition

'making painful decisions' by creating a strong divide between city and countryside. It also represents the three design propositions by utilizing the small scale as a flexible unit within the larger frame of the polders and designing the soil water system in interdisciplinary fashion.

The results of the study were published in several magazines and newspapers, resulting in interest from the local political party *OpenGroenProgressief* (OGP) in the municipality of Midden-Delfland. It aims to make its party program inclusive by organizing thematic discussions in community buildings about important societal topics, in order to involve citizens in thinking how society can respond to big challenges. OGP dedicated one meeting to the importance of water management in Midden-Delfland under the title 'Pump or drown?' The event aimed to engage citizens in discussions about critical issues such as soil subsidence, rising sea levels and increasingly extreme weather patterns. The goal was to involve the community in forward-thinking conversations about how to keep Midden-Delfland safe, liveable and accessible in the future (OGP, 2023).

The party believes that by bringing together people with diverse knowledge and perspectives, the best ideas can emerge. To this end, it invited the RDD team working in Midden-Delfland, known for innovative and sometimes extreme design concepts, which can help inspire realistic solutions. Additionally, three local experts were invited to contribute their insights:

- Marcel Vissers from the Water Board Delfland, who provided expertise on local water management.
- Arie van den Berg, a farmer with extensive experience in farming in harmony with nature, offering deep knowledge about the agricultural possibilities and limitations in the area.
- Lobke Zandstra from the Water Board Delfland, who focused on the preservation and expansion of recreational water use.

Following a presentation by the RDD team, the evening continued with smaller group discussions, where participants could share their knowledge and ideas. These discussions were designed to foster dialogue about the future of Midden-Delfland, allowing residents to voice their concerns, wishes and ideas.

While the RDD research primarily focuses on long-term solutions, which can be challenging for citizens to relate to, OGP chairman Jakob Jongsma identified several actionable steps that residents can begin working on immediately. These include greening gardens to improve local biodiversity and manage rainwater more effectively, replacing solid tiles in parking spaces with open tiles to enhance water infiltration and reduce runoff, and collecting and using rainwater during dry periods instead of draining it through the sewer system. These practical measures can help residents contribute to the resilience and sustainability of Midden-Delfland, aligning with the broader themes discussed during the event.

Following this event, the ZUS and Flux teams continued their collaboration with the Water Board of Delfland to expand and scale the ideas of their initial plan. Their goal was to apply these concepts to a broader area within the Water Board's jurisdiction, with a particular focus on addressing the issue of freshwater shortages. ZUS took the initiative further by establishing a local lab in a farming area within the region. This lab serves as a space where ongoing dialogue between designers and the community can take place, ensuring that the conversation continues beyond the initial event. By fostering this continuous exchange, the lab aims to create a platform for co-creating solutions that are grounded in both expert knowledge and local insights, contributing to the long-term resilience and sustainability of the area.

Future of Everyday Life in the Delta

The RDD study adopts a dynamic, longue durée perspective (Braudel, 1949) on the Dutch delta system, contrasting with the top-down approach in Deltares' solutions, which are based on system characteristics and focus on four scenarios: protected-open, protected-closed, seaward and flexibility (Haasnoot et al., 2019). Deltares explores these extremes to define the playing field.

In the protected-closed scenario, hard or soft protective measures – such as water defences, sand nourishment or wetlands – are implemented, and river arms are closed with dams. The protected-open scenario uses similar protective measures but maintains open river connections to the sea. The seaward scenario creates new, elevated land to shield the delta from flooding, while flexibility reduces vulnerability to sea level rise through adaptive strategies, including salt- or water-tolerant land use (e.g., floating buildings or infrastructure on piles), raising land, spatial planning and relocation.

However, the Deltares study lacks a longue durée perspective, which considers the long-term dynamics of the water-soil system in shaping these scenarios. Instead, its perspectives are purely future-oriented. The longue durée approach (Bloch, 1935; Febvre, 1935; Braudel, 1949) encompasses a tripartite system of short-term *événements*, medium-term conjunctures and long-term structural elements. This method, incorporating insights from economic, urban, social and general history, examines cycles and structural factors, such as prevailing attitudes, resistant frameworks and the enduring influence of the natural environment on human activity and communication.

Although limited to the soil-water system and part of standard urban design analysis, the RDD study of five areas – or five 'moments' in the delta – was conducted in alignment with a longue durée approach. This approach considered both temporal and spatial dimensions, selecting locations that represent different qualities and challenges of the delta system. The resulting designs and manifesto propositions align with historical concepts.

The three decision-making propositions of the RDD manifesto – (1) we are here to stay, we take responsibility, and we can change; (2) we need to know more; and (3) we dare to make painful decisions – can be linked to historical approaches described earlier in this paper. These include designing with a vision for the future, anticipating future developments, and integrating them into current planning. For example, the historical approach of designing at different levels – such as situating human settlements on higher ground to avoid flooding – would involve deciding to stop building in deep polders and retreating to higher areas of the country, which would be a painful decision as well.

The design propositions also reflect historical concepts. Proposition 4 – we utilize the delta paradox: regulation within the dynamics – aligns with the historical approach of protection at local and regional scales, which involved creating socially cohesive dike rings tailored to specific communities. This remains relevant in contemporary Dutch society. For example, in Midden-Delfland, residents and farmers recognize the need to adapt to changing conditions, acknowledging that the engineered landscape is no longer sustainable and that they must actively participate in solutions. Propositions 5 – we design the delta bottom-up – and 6 – the design of the delta starts with the section – correspond to the historical approach of working with the natural system, respecting natural boundaries and the environment's carrying capacities.

The past, present and future of deltas are deeply interconnected through the dynamics of the water-soil system and human occupation systems that interact with it. Awareness of historical approaches is crucial for future delta management. Understanding how we arrived at the current situation requires unravelling how the delta was shaped over time and exploring how we can reassemble it in harmony with nature. Citizens need to be a part of this, because it is a system change that will affect everyday life in the delta.

Figure 1: Living with the dynamics of nature and water. Redesigning deltas design study. Zus, Flux and Sweco, 2022.

References

Bento, A. M., Miller, N., Mookerjee, M., & Severnini, E. (2023). 'A unifying approach to measuring climate change impacts and adaptation', *Journal of Environmental Economics and Management, 121*, 102843. https://doi.org/10.1016/j.jeem.2023.102843

Bloch, M. (1935). 'Avènement et conquêtes du moulin a eau', *Annales d'histoire économique et sociale*. 36 (7), 538-563. (Cambridge, MA: Cambridge University Press).

Braudel, F. (1990) [1949]. *La Méditerranée et le monde méditerranéen à l'époque de Philippe II*, 9th ed. (Paris: A. Colin). Translated by S. Reynolds (1995) [1972]. *The Mediterranean and the Mediterranean World in the Age of Philip II*, 15th ed. (New York, NJ: Harper Collins).

Burke, G. L. (1956). *The Making of Dutch Towns: A Study in Urban Development from the Tenth to the Seventeenth Centuries* (London: Cleaver-Hume).

De Hoog, M., Sijmons, D., & Verschuuren, S. (1998). 'Herontwerp van het Laagland', in J. Blom (ed.). *Het Metropolitane Debat*, 74-78 (Bussum: THOTH).

Haasnoot, M., Diermanse, F., Kwadijk, J., De Winter, R., & Winter, G. (2019). *Strategieën voor adaptatie aan hoge en versnelde zeespiegelstijging: Een verkenning* (Delft: Deltares).

H+N+S Landschapsarchitecten (2023). *De Lagenbenadering, het Lagenmodel en het Casco Concept voor Redesigning Deltas: Reflecties op drie instrumenten in het herontwerpen van de Nederlandse delta* (online). Available: https://hnsland.nl/documents/75/20230301_HNS_

drie_instrumenten_voor_Redesigning_Deltas.pdf (accessed 18 January 2025).
Hooimeijer, F. L. (2014). *The Making of Polder Cities: A Fine Dutch Tradition* (Rotterdam: Jap Sam Publishers).
Hooimeijer, F. L. (2023). *Evaluation Redesigning Deltas Design Study* (Delft: TU Delft/Deltares).
Febvre, L. (1935). 'Réflexions sur l'histoire des techniques', *Annales d'histoire économique et sociale*. 36(7), 538-563. (Cambridge, MA: Cambridge University Press).
National Research Council. (2013). *Delta Waters: Research to Support Integrated Water and Environmental Management in the Lower Mississippi River* (Washington, DC: The National Academies Press).
OGP (2023). *Verslag thema-avond Pompen of verzuipen!?* (online). Available: https://ogp-md.nl/verslag-thema-avond-pompen-of-verzuipen/ (accessed 17 January 2025).
Sijmons, D. F. (1991). *Het casco-concept: een benaderingswijze voor de landschapsplanning* (Den Haag: Informatie-en Kenniscentrum NBLF).
Van Dam, P. J. E. M. (2016). 'An Amphibious Culture: Coping with floods in the Netherlands' in P. Coates, D. Moon, P. Warde (eds.). *Local Places, Global Processes*, 78-93. (Oxford: Oxbow Books).
Van der Ham, W. (2002). 'De historie', in M. van der Buuren, C. van Drimmelen & W. Iedema (eds.), *Waterlandschappen: De cultuurhistorie van de toekomst als opgave voor het waterbeheer*, 107-117 (Lelystad: Ministerie V&W/RIZA).
Van der Woud, A. (1987). *Het lege land, de ruimtelijke orde van Nederland 1798-1848* (Amsterdam: Meulenhof).

Should We Even Live Here?

Gaby S. Langendijk

Is this land meant for water or for humans? Should we even live here? The Netherlands. The Dutch identity and culture have been defined by centuries of struggle against water. While trust in the country's advanced water management systems keeps the perception of risk relatively low, climate change is increasingly challenging the sustainability of these systems and, consequently, the safety of the Netherlands. As the limits of adaptation approach, four potential strategies for the country's future are explored. Surviving in the Netherlands is not just about engineering solutions and innovation; it also requires a cultural shift – one that acknowledges vulnerability, alters perceptions of risk and embraces loss and mourning. Cultural heritage and the arts offer valuable resources for climate change adaptation and must be considered if the Dutch are to continue living in this land.

Waar eens 't gekrijs der meeuwen	*Where once the cry of the seagulls*
Verstierf aan 't eenzaam strand,	*Perished on the lonely beach,*
Daar schiepen zich de Zeeuwen	*There the Zeeuwen created*
Uit schor en slik hun land;	*From the sludge and mud their land;*
En kwam de stormwind woeden,	*And the storm came raging,*
Hen dreigend met verderf,	*Threatening them with destruction,*
Dan keerden zij de vloeden	*Then they turned back the floods*
Van 't pas gewonnen erf.	*From the newly won land.*

Zeeuws-Vlaams anthem (Pattist & Vreeken, 1917).

The Netherlands: a country where the land fades into the sea. A country where the population has battled the water as long as people can remember. The first verse of the anthem of Zeeuws-Vlaanderen, the most South-Western area of the Netherlands, is one of many lyrics describing how the Dutch fought the water and created land for living during the last hundreds of years. The main battle of the Dutch is one of people against the water, where the Dutch aimed at conquering vast areas. They largely succeeded turning water into land, resulting in over 4000 low-lying reclaimed areas, called *polders* (Steenbergen et al., 2009). Nevertheless, with 26% of the country lying below sea level, the Netherlands remains heavily prone to flooding from both sea and rivers, and the battle is a constant undertaking. An impressive system of dikes, pumps, weirs and storm surge barriers is operated day and night to manage water levels as optimally as possible, to safeguard the populations from flooding. The water management system of the Netherlands had been on the winning end and long-term ceasefires between the Dutch and the water had been called in recent decades. It seemed the time had come to draw up a peace treaty and declare the battle won...

However, no such treaty was ever signed. And while humans have gotten confident in their belief of victory, the water might just be regaining strength in the meantime. With human-induced climate change creeping around the corner, humans might even be lending their opponent a hand. Rising sea levels and increases in heavy rainfall events already lead to intensified flooding and will continue to exacerbate flood events over the course of the coming years and decades. The water is striking back. Climate change-related water challenges form a severe threat to the water management system of the Netherlands. The water is mobilizing itself to take back the reclaimed land. The Dutch, who have existed in a constant battle with the water, are facing unprecedented challenges and are required to step up their game. Ancient questions find their way to the forefront once more, but with a new layer of uncertainty due to climate change. How high should the dikes be to protect the population from flooding? How much water needs to be pumped out of the country? Can the Netherlands remain protected from the sea and river flooding? The underlying question is: can the Dutch sustain their battle with water in the light of climate change? While wondering if the Dutch shall continue the battle, they are ultimately asking themselves: is this land destined for humans or for water? Should we even live here?

Adaptation to Climate Change is Not Limitless

An old legend tells that a little boy was called the 'hero of Haarlem' after he stuck his small finger in a hole in a dike, preventing a flooding disaster in the hinterlands (Schultz, 2019). Dikes serve as a barrier between life and death in the Netherlands, and ensuring their functionality is therefore logically regarded as a heroic act. Since then, the water protection and management systems have become increasingly sophisticated and new technologies and techniques now offer plentiful innovations to safeguard the populations from flooding events. This long-standing experience and continuous innovation has created a feeling of security among the Dutch population, a belief in the ability to protect themselves.

Nevertheless, the ability to adapt to climate change impacts isn't limitless, not even for the Dutch. Coastal systems and urbanized areas at coastlines, such as the areas below sea level in the Netherlands, are particularly susceptible to the so-called limits to adaptation (Leal Filho et al., 2021; Thomas et al., 2021; Martin et al., 2022). Being at the forefront of experiencing sea level rise, these areas only have a limited array of adaptation options to safeguard them from rising sea levels. There is a limit to the number of 'little boys' the Dutch can deploy, or to the height of the dikes. These limits to adaptation are divided into two types. Firstly, so-called 'hard' limits refer to situations in which adaptive actions to avoid intolerable risks are no longer possible. Examples of 'hard' limits are extreme heat unbearable to the human body, or, as in the Netherlands, rising sea levels submerging coastal communities. Secondly, 'soft' limits denote contexts where adaptation options may exist but are not currently available due to, for example, insufficient access to funds, weak governance structures and lack of political will. Soft limits can be overcome through social, institutional or technological innovations and transformation (Thomas et al., 2021).

Hard limits to adaptation are appearing on the horizon for the Netherlands. Though, it is uncertain at which speed and strength these limits will occur, as the rate of climate change depends on our efforts to reduce greenhouse gas emissions still on the rise. Sea level rise projections show possible increases of 0.26 m to 1.24 m by 2100 (compared to 1991-2020), or up to 2.5 m if highly uncertain and extreme melting events are considered

(KNMI, 2023). Under these uncertain circumstances, a diversity of strategies can help in adequately planning and implementing adaptation action. Developing a range of strategies can help to understand options, as well as to change course under unfolding climate change impacts over time. A few key strategies are considered for the Netherlands, ranging from fully protecting the country to accommodating sea level rise (Haasnoot et al., 2019). Under a 'protect-closed' strategy the Netherlands is enclosed by tidal barriers and a plentiful number of pumps are employed to manage the water; it can be seen as a 'fortress Netherlands'. A lighter version of a protected Netherlands can be envisioned in a 'protect-open' strategy, where a myriad of dikes and elevated infrastructure and housing provide dry feet to the Dutch population. A third strategy looks at advancing the country towards the sea in a floating and partly protected manner, so-called 'advance'. Another possibility is to 'accommodate' sea level rise through living with the water and/or moving parts of the population to higher grounds (Haasnoot et al., 2019).

So, the battle with water continues. Our fit for fight is not only determined by the dikes and barriers we can build. Our adaptation to climate change is also limited by the values, perceptions, processes and power structures within society. Whether the various strategies for the future of the Netherlands will be followed through depends heavily on cultural perception, political will and the mentality of the Dutch, among other factors. The values and perceptions of a group of people are shaped by the culture of that community. Hence, if culture is a decisive factor in limits to adaptation, it is also decisive for the ability of communities to adapt their societies to climate change. As culture is a constantly evolving and fluid concept, a deeper dive into Dutch culture is required to understand how it influenced adaptation in the past and how it could influence adaptation in the future (Berger et al., 2023).

Dutch Cultural Identity

Hoe ook door overmacht, nood en zee bedreigd,	*No matter how threatened by force majeur, crisis and sea,*
Oeroude, lieve Friese grond,	*Ancient, dearest Frisian soil,*
Nooit werd die hechte, taaie band verbroken,	*That close, tight bond was never broken,*
Die Friezen aan hun land verbond.	*which bound Frisians to their land.*

Frisian anthem (Halbertsma, 1826).

The Frisians, living in the northern part of the Netherlands, started building dikes around 2000 years ago. These dikes were relatively small at the time, with remains of 70-cm-high dikes found in the Frisian villages Peins and Dongjum (Kranenburg, 2015). Around 1200 CE large-scale dike building started to play a central role in the Netherlands and a new institution was established: the Dutch water boards. Ever since, this oldest democratic institution in the Netherlands has been responsible for water management, governed through an independent voting and tax system (Rijksoverheid, n.d.). The establishment of the institution was essential to the country, because the Dutch had to work together to drain the land, manage water levels and ensure everyone remained safe from flooding. A lot of conversation and time was needed to reach consensus on how to do this. Together with the water boards, the Dutch way of working as a group, called 'polderen', was also founded.

Land gemaakt door mensenhanden, vol vertrouwen en met kracht. Waar de zee werd teruggedrongen die zoveel verschrikking bracht.	*Land created by human hands* *full of confidence and with power.* *Where the sea was pushed back* *that brought so much terror.*

Anthem of Flevoland (Zeiler, 1985).

Land manufacturing, creating land for living: the province Flevoland would not have existed without the Dutch's ability to turn water into land. The word 'maakbaarheid', roughly translatable into 'manufacturability', is at the heart of the deep conviction of the Dutch that they can engineer and innovate a way out of all water-related matters. The saying 'pompen of verzuipen' (pump or drown) originally used in the context of a ship leakage, is applicable to large parts of the Netherlands now. Day in, day out, the pumps are working and the water is pumped from the low-lying areas towards the sea. The second sentence of the anthem of Flevoland, 'full of confidence and with power', illustrates this mentality. Living by and with the sea for so long, the cultural identity of the Dutch is shaped by it. The Dutch have a winner's mentality – a mindset that believes in the power of engineering and innovation, a belief system that is confident that, being Dutch, we can manage the water.

Dutch Delta Roulette

Although aware of its existence, the average modern-day Dutch citizen seems to take the sophisticated water management system as a given, underestimating its fundamental role in the day-to-day existence of the country (Kester, 2024). The Dutch generally believe in a safe future for the country, at least for the foreseeable period. This is characterized by the willingness of citizens to buy houses in the most low-lying areas, such as the famous cheese-making city of Gouda. The region around Gouda is heavily subjected to land subsidence, and is sometimes nicknamed the bathtub of the Netherlands. Despite water levels and unstable soils threatening the region, a new neighbourhood is being constructed in Gouda and citizens have already moved into their homes (Municipality of Gouda, 2024).

Why do people keep on building houses in such areas? Why do we still buy these houses? The perception of risk is at play. Dutch Delta Roulette. A gambling game where we hope everything will turn out to be alright. A game where the Dutch act as this self-assured player who joins the table and goes all-in on water management. With the risk of losing it all. Are we foolishly optimistic, or so smart that we can actually engineer and innovate our way out?

While limits to adaptation are approaching, it becomes harder and harder to protect ourselves from the rising waters. Cultural beliefs and a sense of security are deeply ingrained in the Dutch mindset, but could be slowly losing their legitimacy. What if the Dutch would start seeing themselves as a vulnerable population at risk? Would this change the course of action; would it change the risk perception of the population and thus their behaviour? A mindset shift towards a stronger acknowledgement of vulnerability would certainly require a change in belief systems, which are so closely tied to identity. It would require imagination. What would the lyrics of an anthem then become?

Cultural Heritage for Adaptation

Zuiderzee, Zuiderzee,	*Zuiderzee, Zuiderzee,*
Oude, trouwe, blauwe zee,	*Ancient, loyal, blue sea,*
Je verdwijnt met je wel en wee,	*You will disappear with your good and bad,*
Met je botters en je jollen,	*With your botters and your dinghies,*
Met je harinkies en schollen,	*With your herrings and your plaice,*
Neem je straks ons hart ook mee.	*You will also take our hearts away.*
Zuiderzee.	*Zuiderzee.*

Zuiderzee, excerpt (Davids, 1933).

We can keep on battling and keep on gambling. However, limits to adaptation will arise and the way the Dutch live their lives will change. The vast lands of the Netherlands did not look the same throughout the last centuries. The Zuiderzee, once a wild sea, has been turned into the IJsselmeer after the Afsluitdijk was constructed, tempering the wild sea into a subdued closed-off lake. The song *Zuiderzee* provides a melancholic perspective from fishermen saying goodbye to the sea that disappeared.

De zee ontziet geen mens, geen land;	*The sea spares no man, no land;*
Een stem spreekt, en de dijken breken,	*A voice speaks, and the dikes break,*
En, machtiger dan vele preken,	*And, mightier than many sermons,*
Raakt die stem hart en ingewand:	*Does that voice touch heart and organs:*
Eerst als de kracht ons is bezweken,	*Only when our strength has failed us,*
Is het ons weer een keer gebleken,	*It becomes apparent once more to us,*
Dat wij zijn allen in Gods hand.	*That we are all in God's hand.*

Ballade van den Watersnood/Ballad of the Flood, excerpt (Werumeus Buning, 1970).

Countless songs and poems, stories and monuments, rites and traditions, were developed over time in response to natural hazards, changes in land management or climatic changes (Berger et al., 2023). A distinct subgenre in the Netherlands is centred around disasters, so-called 'rampenliederen' (disaster songs). A large share of these poems and songs are about water-related challenges, such as floods. The *Ballade van den Watersnood* is just one example (Metz, 2022). At the time, these songs helped people to deal with loss and find consolation. In the future we might face loss of land and lives due to intensified flooding and sea level rise once more. This type of cultural heritage could help us cope with climate change impacts; it can support the mourning process.

Culture is a reflection of the values, customs, beliefs and symbolic practices by which men and women live (Berger et al., 2023; Eagleton, 2016). The related production of arts and artefacts within a cultural setting can reorient values and meaning, as well as forms of thinking and feeling, and therefore help shape the identity of a larger group. The Dutch identity of engineers winning the battle with water is under pressure. The arts can stimulate imaginations towards a different identity for the Dutch, for instance one that underlines its vulnerability to water-related challenges. The arts may help reshaping narratives around climate risks and the liveability of the Netherlands, to better understand what is at stake in playing Dutch Delta Roulette. The arts and cultural heritage have the potential to serve as a rich resource for climate adaptation – not solely for communicating climate change, but for reshaping narratives, dealing with loss and mourning, as well as imagining an identity change for Dutch society.

Closure

Is this land destined for water or for humans? Should we even live here? The Dutch identity and culture are shaped by their battle with water during the last centuries. The perception of risk is relatively low, because of a sense of trust in the water management and innovation capabilities of the country. Dutch Delta Roulette seems still a game to play for many. Climate change is increasingly pressuring the Dutch water management system and thus the safety of the country. The limits to adaptation are closing in and four different strategies for the future of the Netherlands are laid out. Continuing to live in the Netherlands is not solely about engineering and innovations; it requires a cultural change which embraces its vulnerability, shifts its risks perception, and can deal with loss and mourning. For this, cultural heritage and the arts are rich resources for climate change adaptation which must be considered if we wish to continue to live here.

Is the Netherlands a logical place to live? No, but we do and we will. The battle with water continues.

Note: This piece reflects the personal opinion of the author and does not express the views or opinions of the affiliated institution. The excerpts of the anthems, poems, and lyrics are freely translated into English by the author of this piece.

References

Berger, H. M., Erbach, T., Kuehn, A., Possman, J. P., & Langendijk, G. S. (2023). 'Rising Waters, Rooted Memories: Cultural Heritage as a Resource for Climate Adaptation in Sinking Cities', *Journal of Delta Urbanism 4*. https://doi.org/10.59490/jdu.4.2023.7334

Davids, L. (1933). *Zuiderzee.* Available: https://www.dbnl.org/tekst/davi013mooi01_01/davi013mooi01_01_0064.php (accessed 18 June 2025).

Eagleton, T. (2016). *Culture* (New Haven, CT: Yale University Press).

Haasnoot, M., Diermanse, F., Kwadijk, J., de Winter, R., & Winter, G. (2019). *Strategieën voor adaptatie aan hoge en versnelde zeespiegelstijging: Een verkenning.* Deltares. Available: publications.deltares.nl/11203724_004.pdf (accessed 18 June 2025).

Halbertsma, E. (1826). *Frisian anthem*. Available: https://historiek.net/volkslied-van-friesland-alde-friezen/83098/ (accessed 18 June 2025).

Kester, J. (2024). *Overheid én burgers zijn volgens de meeste mensen slecht voorbereid op ramp: 'We overschatten onze zelfredzaamheid'.* EenVandaag. Available: https://eenvandaag.avrotros.nl/panels/opiniepanel/alle-uitslagen/item/overheid-en-burgers-zijn-volgens-de-meeste-mensen-slecht-voorbereid-op-ramp-we-overschatten-onze-zelfredzaamheid/ (accessed 18 June 2025).

KNMI. (2023). *KNMI'23 klimaatscenario's voor Nederland*. Available: https://www.knmi.nl/kennis-en-datacentrum/achtergrond/knmi-23-klimaatscenario-s (accessed 18 June 2025).

Kranenburg, R. (2015). *Dijken van Nederland.* Geografie.nl. Available: https://geografie.nl/artikel/dijken-van-nederland (accessed 18 June 2025).

Leal Filho, W., Krishnapillai, M., Sidsaph, H., Nagy, G. J., Luetz, J. M., Dyer, J., Otoara Ha'apio, M., Havea, P. H., Raj, K., Singh, P., Rogers, T., Li, C., Boodhan, M. K., Wolf, F., Yayeh Ayal, D., & Azadi, H. (2021). 'Climate Change Adaptation on Small Island States: An Assessment of Limits and Constraints', *Journal of Marine Science and Engineering*, *9*(6), 602. https://doi.org/10.3390/jmse9060602

Martin, M. A., Boakye, E. A., Boyd, E., Broadgate, W., Bustamante, M., Canadell, J. G., Carr, E. R., Chu, E. K., Cleugh, H., Csevár, S., Daoudy, M., de Bremond, A., Dhimal, M., Ebi, K. L., Edwards, C., Fuss, S., Girardin, M. P., Glavovic, B., Hebden, S., ... Zhao, Z. J. (2022). 'Ten New Insights in Climate Science 2022', *Global Sustainability*, *5*, e20. https://doi.org/10.1017/sus.2022.17

Metz, T. (2022). *Rampliederen helpen ons overstromingen te verwerken en te herdenken.* Nattigheid.nl. Available: https://www.nattigheid.nl/rampliederen-om-overstromingen-te-verwerken-en-te-herdenken/ (accessed 18 June 2025).

Municipality of Gouda. (2024). *Nieuwbouwwijk Westergouwe*. Available: https://www.gouda.nl/gouda-werkt-aan-de-stad/nieuwbouwwijk-westergouwe/ (accessed 18 June 2025).

Pattist, J. N. & Vreeken, J. (1917). *Zeeuws-Vlaams volkslied*. Available: https://www.zeeuwsenzo.nl/tekst-zeeuws-vlaams-volkslied/ (accessed 18 June 2025).

Rijksoverheid. (n.d.) *Waterschappen.* Available: https://www.overheid.nl/wie-vormen-de-overheid/waterschappen (accessed 18 June 2025).

Schultz, B. (2019). 'The Story of the Dutch Boy Who Prevented a Flooding Disaster: Origin and Variations on the Theme', *Water History*, *11*(3-4), 207-231. https://doi.org/10.1007/s12685-019-00237-0

Steenbergen, C., Reh, W., Nijhuis, S., & Pouderoijen, M. (2009). *De Polderatlas van Nederland – Pantheon der Lage Landen* (Bussum: Thoth).

Thomas, A., Theokritoff, E., Lesnikowski, A., Reckien, D., Jagannathan, K., Cremades, R., Campbell, D., Joe, E. T., Sitati, A., Singh, C., Segnon, A. C., Pentz, B., Musah-Surugu, J. I., Mullin, C. A., Mach, K. J., Gichuki, L., Galappaththi, E., Chalastani, V. I., Ajibade, I., ... Bowen, K. (2021). 'Global Evidence of Constraints and Limits to Human Adaptation', *Regional Environmental Change*, *21*(3), 85. https://doi.org/10.1007/s10113-021-01808-9

Werumeus Buning, J. W. F. (1970). *Verzamelde gedichten* (Amsterdam: Querido). Available: https://www.dbnl.org/tekst/weru002verz03_01/weru002verz03_01_0145.php (accessed 18 June 2025).

Zeiler, M. (1985). *Anthem of Flevoland.* Available: https://www.flevoland.nl/wie-zijn-we/over-flevoland/volkslied-van-flevoland (accessed 18 June 2025).

Climate & Health: A Shared Prescription for a Better Future

João Cortesão and Agnès Patuano

Urban climate resilience and public health are inextricably connected. While the body of knowledge on these two topics is growing, little attention has been paid to their intersections. In this article we plead for thinking of urban climate resilience and public health as integrated concepts, as this can lead to more complete and, thus, more beneficial solutions for society and space. After introducing the context behind this argument, we present a brief theoretical perspective on urban climate resilience and public health explaining why integrated approaches to both matter. The article then presents projects and studies that illustrate what that integration can mean in practice. The examples we present come from the academic (Wageningen University & Research) and business (Sweco Netherlands) sectors and refer to both practical engineering and policy topics. We then delve deeper into the integration of climate and health by mapping their intersections and identifying the trade-offs inherent to these intersections. We conclude with the foreseeable impacts of the integration of climate and health on society and space, in delta regions and elsewhere.

Context

Urban climate resilience and public health are cornerstones of urban resilience. The former relates to spatial measures that (re) shape urban areas towards coping with the impacts of extreme weather, for example heat waves or flooding. The latter deals with protecting and improving the quality of life and wellbeing of the people living in these areas, for instance, by providing better opportunities for physical activity, relaxation or social connections. These topics are interconnected: health depends largely on the physical context one lives in; urban climate resilience, albeit of a design and engineering nature, is ultimately about securing everyday life and improving the health and wellbeing of urban residents. It is precisely this interconnection that renders both climate and health core assets of urban resilience. A better understanding of these intersections can lead to more systemic, integrated (holistic) resilience approaches.

While there is solid knowledge on the topics of urban climate resilience and public health, their intersections and trade-offs are underexplored. The few studies that associate these topics tend to go in depth on one topic while addressing the other at a more superficial level. For example, studies may indicate a given heat stress reduction by increasing the urban tree canopy without referring to the potential effects of that intervention on pollen allergies or mental health. Integrated approaches combining climatic, natural, ecological and urban planning considerations with social and health considerations are vital to address the complex relationships between these topics. These approaches allow for a more holistic understanding of how climate and health are interconnected and enable more effective, long-term solutions to be developed and implemented. This can be of profound influence in delta regions, where the impact of rising sea levels, climate change, and human activities can significantly affect land use, time, and daily life.

Fuelled by these concerns, this article expands on the need for thinking of urban climate resilience and public health as integrated concepts, as this can lead to the integrated spatial solutions that urban resilience calls for. We map the intersections between urban climate resilience and public health – namely between key climate risks, urban climate resilience measures and health benefits – along with the trade-offs inherent to these intersections. In doing so, this article builds on existing knowledge to unveil pathways towards integrated approaches to climate and health. Thus, we bridge science and practice to propose a perspective for addressing the climate crisis and its effects on health in urban areas.

Climate and Health in Theory

As concepts, urban climate resilience and health have much in common. Both span a wide range of scales, attributes and factors. They are both used as umbrella terms to express different phenomena, which depend both on internal characteristics of individuals or cities (such as genetic predispositions or urban structures), and on external factors (such as pollution or unpredictable weather events). Urban climate resilience and health are complex concepts that we define for the sake of clarity as:

> Urban climate resilience: the capacity of an urban area to maintain or quickly return to desired functions in the face of climate-related chronic stresses and/or acute shocks, to adapt to changing and uncertain climatic conditions, or to rapidly transform in ways that build climate-adaptive capacity across temporal and spatial scales. The result of infrastructural measures ... with the ultimate purposes of preserving infrastructure and increasing the capacity of urban populations to cope with the impacts of climate change in ways that maintain a shared urban identity (Cortesão & Copeland, 2021, 1).
>
> Health: a state of complete physical, mental and social wellbeing and not merely the absence of disease or infirmity (World Health Organization, 1948).

We can trace back studies on urban climate resilience over several decades, within the scope of spatial design and engineering. Initially used for military and aerospace applications, the 1960s shed light on the importance of 'thermal comfort' (Jones, 2001). The following decades saw the rise of key concepts such as 'sustainable urban development', 'urban regeneration' (Couch et al., 2011), the 'sustainable city' (Jenks et al., 1996) or 'green urbanism' (Calthorpe, 2011). More recently, 'climate-responsive design' and synonyms such as 'bioclimatic design' or 'climate-conscious design' have been providing an array of spatial measures for various climatic regions, with the Netherlands playing a particularly fruitful role (Kleerekoper, 2016; Tammingaet al., 2020; Kluck et al., 2021; Brandsma et al., 2024).

Meanwhile, the effect of the built environment on health has long been recognized, with its significance dating back to the moment John Snow identified a single contaminated water pump as the source of a severe cholera outbreak in London in 1854 (Cameron & Jones, 1983). This groundbreaking discovery not only marked the beginning of modern epidemiology but also highlighted how urban planning and infrastructure design could directly influence public health outcomes (Corburn, 2004). Since then, urban planning and design have evolved with a growing awareness of the need to build healthier cities, mitigating the harmful effects of rapid industrialization by considering factors like sanitation, housing quality, and access to clean water and air (Corburn, 2004). Recently, the scope of this relationship has expanded beyond just physical health to include mental health. Indeed, the built environment impacts overall wellbeing (Barton, 2005; Mouratidis, 2021). Factors such as noise pollution, access to nature, walkability, and social interaction opportunities all contribute to mental health and quality of life (Mouratidis, 2021). This broader understanding is reflected in the growing popularity of the One Health approach, which further strengthens the connections between human, animal, and environmental health (World Health Organization, 2024). Although there is no universally agreed-upon definition, One Health acknowledges that human health is deeply intertwined with the health of other species and ecosystems. As our understanding of health evolves, this approach emphasizes the need for integrated strategies that consider not just the prevention of diseases but also the promotion of physical and mental wellbeing in everyday urban and natural environments alike.

The intersections between the two concepts are not obvious: although the ultimate goal of urban climate resilience is to contribute to health and wellbeing in urban areas, it operates mostly at the biophysical level of the built environment, while health is still primarily a human-centred topic. However, these intersections do exist. Both urban climate resilience and health exist at the perceptual (experiential) and physical (factual) levels.

For example, climate and atmospheric conditions can impact our perceptions in quite different ways. The physical experience(s) of an environment and the health outcomes it leads to can depend on palpable spatial characteristics, such as paving materials or number of trees, as much as they depend on individuals' specific health conditions or needs. Visiting green spaces seems to be more beneficial to people from lower socio-economical classes (Mitchell & Popham, 2008). Likewise, the impacts of extreme weather events depend on spatial elements such as flood defences or the orientation of building blocks. But the perceptual experience depends on personal factors such as age, gender, body mass or previous thermal experiences (Nikolopoulou et al., 2001; Tseliou et al., 2010; Lenzholzer & Nikolopoulou, 2020). Bottom line: physical and perceptual experiences might not necessarily match. That is why, for example, the temperature one actually feels outside is a more useful indicator than air or surface temperatures when it comes to estimating the impacts of heat stress on public health.

Integrated approaches to urban climate resilience and public health are thus topics subjected to uncertainty and variability. The healthy and safe everyday life of urban populations in delta regions and elsewhere is strongly dependent on this integrated understanding. But where does biophysical urban climate resilience intersect with human-centred public health? And what are the trade-offs inherent to these intersections?

Climate and Health in Practice

Let us start answering these questions by discussing some projects and studies that illustrate what this integration can mean in practice. These examples are diverse as they were conducted within the academic (Wageningen University & Research (WUR)) and the business (Sweco Netherlands) sectors and refer to both practical engineering and policy topics: the practice-oriented research GreenQuays (WUR); the research project 'Blue and green infrastructure to beat the urban heat – BENIGN' (WUR); the white paper 'Voorkomen is beter dan genezen' ('Prevention is better than cure'; Sweco, 2022); and the policy-oriented research 'Een praktische verkenning naar een Nederlandse groennorm' ('A practical exploration towards a Dutch green standard'; Sweco, 2024). We briefly describe these projects and highlight their lessons in relation to the integration of climate and health.

GreenQuays

GreenQuays was a project co-financed by the European Regional Development Fund through the Urban Innovation Actions Initiative. The project gathered partners from different disciplines and sectors engaged into delivering a systemic approach to climate and social resilience. Partners on the academic side included WUR and TU Delft, while the practice side was composed of the City of Breda, the Netherlands, and its design team, five regional flora and fauna associations and one plant nursery. Transdisciplinary and cross-sectoral practices were key for coupling climate resilience with nature-inclusiveness, feasibility and the municipality's vision for the site.

The project gave back the river Mark to the city of Breda by daylighting its buried segment and refurbishing its open-air stretch (Figure 1). The analysis of the site revealed its vulnerability to urban heat stress and pluvial flooding, which led to the underuse of the public spaces adjacent to the river and a lack of significance in people's everyday life. This was mostly due, climate-wise, to the excessive exposure to direct solar radiation combined with predominantly hard-paved outdoor spaces with little or no vegetation (Cortesão & Forgaci, 2023). To address these challenges, GreenQuays developed a network of self-sustaining green or soft-paved spaces characterized by the increase of shade and evapotranspiration, local rainwater retention and through the 'Nature-Inclusive Quay' technology, that supports the natural growth of small plants, mosses and trees in the quay walls (Mulder et al, 2023).

Integration of climate and health

The GreenQuays project comprised the improvement of the health and wellbeing of local populations through the provision of more and better-quality public spaces, more vegetation, more contact with the water and all the physical, mental and social benefits resulting from looking at and being amidst green and blue spaces. For instance, GreenQuays applied knowledge on the cooling effects of small urban water bodies (Jacobs et al., 2020), such as increasing shade along the water body or enabling more access to water. The climate-resilient design measures were targeted at people, at improving the conditions offered for better health, wellbeing and social interaction. While these measures were thought at the biophysical level of the built environment, their aim was human-centred: physiological (less heat stress on the body) and psychological cooling (the sensory experience of shaded, green water environments). The ultimate effects were improved recreation and social interaction and cohesion.

BENIGN

The BluE and greeN Infrastructure desiGned to beat the urbaN heat (BENIGN) project is an ongoing research funded by the Dutch Research Council (NWO) that aims to investigate how blue and green infrastructure (Figure 2) can be employed in urban areas to create healthy living conditions. Healthy living conditions are defined as the results of indoor and outdoor climate on physical and mental health. This can be seen in the effects of heat stress, changes in water quality and plant diversity on health. In this project, urban living conditions and their effect on citizens' health are investigated. BENIGN employs an integrated approach by combining the disciplines of health with thermophysiology, water and plant ecology, spatial design, spatial planning and governance, behavioural and cognitive science and geo-information science. This multidisciplinary approach is ensured by members of different parts of the knowledge chain: fundamental (knowledge institutions), applied and practice-oriented (companies) and civil society organizations (municipalities, citizens and civil organizations).

The end goal is to synthesize the project findings into a decision support system for municipalities to assist them in creating healthier urban living conditions. To achieve this outcome, the project includes real-life physical and social experiments in the built environment by co-designing and implementing blue and green interventions in three living labs in three municipalities in the Netherlands.

Integration of climate and health

In this project, climate and health are integrated through three main pathways considering the effects of heat on: 1) water quality, 2) pollen allergies, and 3) vulnerable urban residents' perception and use of green and blue spaces during heatwave periods. The starting point of the project was the acknowledgement of the intersections between urban heat and health issues, the belief that these issues are best addressed with the implementation of green and blue climate-resilient infrastructure, and by deeply considering the health outcomes resulting from this infrastructure. The clearest example of how climate and health are embraced as one in this project is that the climate resilience spatial measures proposed by the project will be pay specific attention to the health and wellbeing of people with chronic health conditions, such as asthma or cardiovascular disease, who are amongst the most vulnerable during extreme heat periods. Taking pollen allergies into account adds a layer of complexity to the employment of vegetation in the climate resilience measures developed. This needs to be reasoned bearing in mind not only the cooling effects they lead to, but also the utmost minimization and prevention of release of allergenic substances.

'Voorkomen is beter dan genezen'

The white paper 'Voorkomen is beter dan genezen' ('Prevention is better than cure'; Sweco, 2022) connects scientific knowledge and practice on the increase of vegetation in urban areas (urban greening) with the physical and social domains, where health and well-being is a core component. The paper is a 'manifesto' on how urban greening with climate concerns is essential for preventing health hazards in cities, in particular related to heat stress. Drawing on scientific knowledge, a four-steps plan is outlined for Dutch municipalities to plan and implement smart, fair and healthy urban greening strategies (Figure 3). The paper also includes a calculation of the financial benefits of urban greening, based on scientific insights and using neighbourhood data from 104 Dutch municipalities.

Integration of climate and health

Sweco's white paper confirms that increasing vegetation in the built environment can positively impact health and wellbeing by reducing the vulnerability of people to extreme climate conditions (Sweco, 2022). By doing so, it supports the conception of climate and health as integrated concepts. The step-by-step plan offers municipalities guidance to capitalize their greening potential (that is, surface area available for increasing vegetation) in a way that holds the best health impacts. Sweco's analysis of different Dutch neighbourhoods shows that those where the most health benefits can be achieved through urban greening are often socio-economically vulnerable neighbourhoods. This is of particular importance as lower income residents are amongst the most vulnerable groups when it comes to coping with the effects of climate change.

Regarding costs, the paper shows that every hectare of extra urban greenery means, on average, more than eight fewer residents requiring health care during heat stress periods. This represents significant financial savings in social costs (especially associated with hospital admissions and death tolls) both for municipalities and for other parties within and outside the healthcare sphere. The savings amount to approximately 125 million euros per year for all municipalities analyzed. The paper argues that if the greening actions applied in these municipalities were to be extended to all Dutch municipalities, annual savings could reach 400 million euros. The paper emphasizes that urban greening targeted at improving health transcends the spatial domain and requires innovative collaboration and financing mechanisms.

'Een praktische verkenning naar een Nederlandse groennorm'
The policy-oriented research 'Een praktische verkenning naar een Nederlandse groennorm' ('A practical exploration into a Dutch green standard'; Sweco, 2024) was set up to inform the establishment of a green standard for the Netherlands. The motivation behind this research is that green standards can back up the incrementation of vegetation in cities (Figure 4) and, thereby, help address the climate and health challenges brought on by climate change. A national green standard sensitive to local circumstances has the power to accelerate change in the deltaic Dutch landscape towards climate resilience by providing guidelines for decision-makers, spatial designers and planners and developers.

Sweco conducted this research anchored by the question: is there already a good green standard that can be used across the Netherlands? To answer this, scientific knowledge was combined with practical experience on the topic among municipalities, provinces and developers and data from the Groene Stad Challenge. The focus of the study were ten existing Dutch and international green standards well-known in the Netherlands. Not all documents were green standards per se, but they were widely recognized reliable sources of information which, fully or in part, form the backbone of urban greening in the Netherlands. The standards were assessed and compared to one another, which resulted in a list of top 3 best-rated standards. Based on the conclusions of the assessment process, the research provides five recommendations for a green standard applicable to the Netherlands.

Integration of climate and health
This research concluded that, as of yet, there is no good green standard in the Netherlands. The main reason is that none of the examined standards fully addressed important policy themes around urban greening: climate, health and biodiversity. Following urban resilience thinking, a good green standard should be

integral and provide guidelines on these themes. The examined standards mostly focussed on one theme, or unequally addressed themes when they (seldom) combined them. For example, a guideline about increasing shade to combat heat stress would not specify or mention the benefits this entails to health. Likewise, a guideline stating the importance of movement outdoors would not refer to the importance of shade in allowing outdoor activity to happen. This research therefore recommends establishing an integral green standard, that is, a standard that can address extreme climatic phenomena, as well as improve health and support urban biodiversity. The integration of climate and health is, therefore, a core principle in this research.

Proposal for Integrating Climate and Health

Intersections

Now that the integration of climate and health have been sketched both theoretically and practically, we propose an alternative definition of urban resilience. Urban resilience can be defined as 'the ability of a socio-technical-environmental system to sustain, improve and innovate its key functions – through absorbing, reacting to, recovering from, adapting to or reorganizing – in response to chronic stresses, abrupt shocks, and disruptions' (DeSIRE, 2020).

Based on this definition, we propose the following: urban resilience is the capacity of an urban area and all its residents to maintain or quickly return to desired states and functions in the face of chronic stresses, acute shocks and/or disruptions, to adapt to changing and uncertain conditions, or to rapidly transform in order to preserve its infrastructure and a state of complete physical, mental and social wellbeing, with a minimum amount of damage and losses.

Above, we raised the question: where does biophysical urban climate resilience intersect with human-centred public health? With our definition of urban resilience in mind, and based on the literature listed in the bibliography, we propose answering this question with Figure 5. This figure depicts an overview of the main intersections between urban climate resilience and public health. These intersections are 'mapped' taking into account: 1) key climate risks in urban areas like heat stress, pluvial flooding, (flash) flooding and drought; 2) the main urban climate resilience measures that address those risks, where one measure may address more than one risk; and 3) the main health benefits resulting from these measures. Soil subsidence, a key climate risk in delta areas, was excluded from Figure 5 as it is often entangled with droughts or floods, and therefore relates to similar urban climate resilience measures and benefits to public health.

The intersections depicted in Figure 5 vary and can relate

to multiple key urban climate risks and health benefits resulting from urban climate resilience measures. In other words, a benefit to health might relate to more than one measure and key climate risk. Furthermore, all intersections are bilateral: urban climate resilience influences health; health is the fundamental goal of urban climate resilience. We believe it is critical to acknowledge and understand these intersections, as well as to proactively implement measures that target mitigating the impacts of climate change intrinsically support health in urban areas. These measures can be both infrastructural and managerial; the former relate to spatial design and engineering solutions, while the latter relates to, for example, adapting healthcare systems to be better prepared for health risks associated with the impacts of climate change.

Recent integrated approaches to health, such as One Health, broaden our understanding of health and encourage us to work in more inter- and transdisciplinary ways, in order to propose more effective and impactful solutions that foster both climate resilience and health. Embracing the presented intersections is a collective effort that requires the involvement of a broad spectrum of stakeholders including spatial planners, public health professionals, environmental scientists, policymakers, community organizations and society generally.

Relevance to Delta Regions

Delta regions are a good example of why the integration of urban climate resilience and public health is of critical importance: they are particularly vulnerable to the impacts of climate change due to their low-lying geography and their dependence on water resources. One of the main risks is flooding, either through extreme weather events, like storms or heavy rainfall, or as a result of sea-level rise.

Floods cause injuries and drownings and spread waterborne diseases (Agache et al., 2022). The contamination of water sources with sewage, industrial waste, and agricultural runoff can introduce pathogens such as bacteria, viruses, and parasites into the water supply (Agache et al., 2022; Ahern et al., 2005; Paterson et al., 2018). Floods caused by sea-level rise create saltwater intrusion into freshwater sources which threatens drinking water supplies and agricultural land (Schmidt, 2015).

Flooding can also damage infrastructure, affecting access to healthcare services and critical resources like electricity, food and clean water. Floods cause material damages to individuals and their homes which might force them to relocate. The displacement of populations creates its own health risks, as affected populations often face overcrowded living conditions in temporary shelters or informal settlements. Such circumstances can increase the risk of disease transmission due to poor sanitation and limited access to clean water and healthcare services (Paterson et al., 2018). Furthermore, the loss of one's home has been associated with elevated levels of anxiety, depression, and post-traumatic stress disorder (Ahern et al., 2005; Paterson et al., 2018). Floods also have an impact on erosion and soil stability, which can further damage infrastructure.

The intersections identified in Figure 5 for 'pluvial flooding' and '(flash)flooding' are the most relevant to delta regions as they relate to the main climate risk there: flooding. Although we will not discuss this for conciseness, we acknowledge that the remaining risks are also of importance to delta regions.

Trade-offs

Above, we also raised the question: what are the inherent trade-offs of the intersections of climate and health? Our answer is that although we see these intersections as key for the future of society and space, they do entail trade-offs which cannot be overlooked. Here, we highlight some critical points:

The mismatch between benefits to climate and benefits to health. An urban climate resilience measure good at improving urban microclimates might impoverish the conditions offered for health. For example, employing minimal paving or soft pavers are efficient solutions for decreasing surface and air temperatures (coping with heat stress), allowing rainwater to infiltrate locally (coping with pluvial flooding) or for allowing to store rainwater underground (coping with drought). Depaving also often involves community engagement and participation, providing opportunities for residents to come together to improve their neighbourhoods (social health) (Tamminga et al., 2020). Spaces created through such initiatives can serve as gathering places for social interaction, recreation, and relaxation, contributing to overall community wellbeing. However, removing hard surfaces might also limit accessibility to more vulnerable populations, such as elderly people or those with reduced mobility. This can have negative consequences for their ability to access vital resources but also increase their risk of social isolation, leading to loneliness and depression. So, whereas this type of measure might be effective climate-wise, that might not be the case health-wise. Likewise, an effective health-focused design like improving accessibility, might not be effective climate-wise, for instance, if there is no compromise between paved (for accessibility) and depaved (for example, for local water infiltration) ground surfaces. Dialogue with local communities combined with smart design choices might help prevent the negative impacts of this trade-off.

The risk for urban greening to promote social injustice. Increasing green spaces and vegetation in urban areas can significantly contribute to coping with extreme weather (for instance, heatwaves), to improve air quality and to enhance overall wellbeing. But urban greening can also contribute to rising property values and thus to displacing low-income residents (gentrification), as vegetation makes neighbourhoods more desirable. Thus, while urban greening is one of the most effective measures for delivering benefits for both urban climate resilience and public health, careful planning and community engagement are essential to mitigate the risk of exacerbating socioeconomic inequalities and ensure that these benefits are equitably distributed among all residents (Wolch et al., 2014).

Plants and allergies. While increasing vegetation in urban areas is, climate-wise, the measure holding the most benefits, some species might release pollen that can exacerbate allergy complaints in urban areas. Furthermore, higher temperatures mean the flowering season of grasses, trees and herbs begins earlier and lasts longer. Consequently, more people are developing pollen allergies (Damialis et al., 2019). This should be considered when deciding on the type and characteristics of the plants chosen to be implemented in an urban area. A different weighing between the use of plant types is necessary. Hiemstra (2018) and the Dutch organization Natuur & Milieu (2023) have published databases that can assist with this endeavour.

Investment and collaboration. Implementing measures that integrate climate and health requires sufficient funding and calls for engaging a variety of stakeholders. Funding might not always be available, or it might not prioritize implementing these measures. Next to this, collaboration requires coordinating stakeholders with different expertise, priorities and objectives, which can be challenging. Such challenges may require bridging gaps the between academic disciplines and fostering effective communication and cooperation amongst parties. Channelling funding and ensuring positive collaboration among stakeholders takes considerable time and dedication. At the same time, the climate and health crises call for urgent and decisive action.

Need for space. Implementing measures that integrate climate and health requires sufficient space, which might not be available. This is particularly challenging in urban contexts where spatial pressures are high (for example, downtown districts). These pressures can relate to lack of room in the streetscape, buildings, underground and overground infrastructure, or to natural circumstances such as water bodies in delta regions.

Conclusion

This article is a plea for integrated solutions addressing two critical challenges for the future of society and space: climate and health. Our aim is to contribute to developing an unexplored approach to urban resilience for delta regions and elsewhere. We support the perspective that new urban developments should be based on the integration of urban climate resilience and public health. We mapped the main intersections between the two fields as we believe that acknowledging these intersections is vital for rendering urban areas and societies in delta regions (and elsewhere) future-proof. Without urban climate resilience we cannot address the climate crisis. But climate-resilient design solutions are a means to achieve the goal, not the goal itself. The goal is the health and wellbeing of urban populations, both present and future ones, in delta regions and beyond. Without healthy people outdoors, the public realm loses meaning and so does, thus, the purpose of spatial design.

Acknowledging and understanding the intersections and trade-offs between climate and health illustrates how the traditional hard engineering approach is evolving, thanks to transdisciplinary and cross-sectoral research and practice. The projects and studies presented above illustrate the importance of transdisciplinary research and emphasize the growing need to implement scientific findings into tangible urban resilience actions. Scientific and business partners can and must support climate transition efforts. The examples presented above show how, for example, WUR is doing this by investing in practice-oriented and open access research, and Sweco is translating scientific insights into concrete recommendations for policy and urban projects.

We would like to finish with highlighting the red thread of our 'prescription': acknowledging and implementing the integration of climate and health represents a leap forward in urban resilience knowledge and practice. This has two inextricably connected beneficiaries: our urban areas and our societies. Be it in delta regions or be it elsewhere.

Figure 1: View of the river Mark after refurbishment during the GreenQuays project, Breda. Photo: Veerle Hemerik, 2024.

Figure 2: Blue and green infrastructure can help create healthy urban environments for everyday life, Utrecht. Photo: João Cortesão, 2024.

Figure 3: An urban green space comprising smart, fair and healthy urban greening strategies, Parc de Bruxelles. Photo: João Cortesão, 2024.

Figure 4: Example of urban greening by the incrementation of vegetation in urban areas, at both the public space and building levels. These actions can excel when backed up by green standards. Tokyo, Japan. Photo: João Cortesão, 2023.

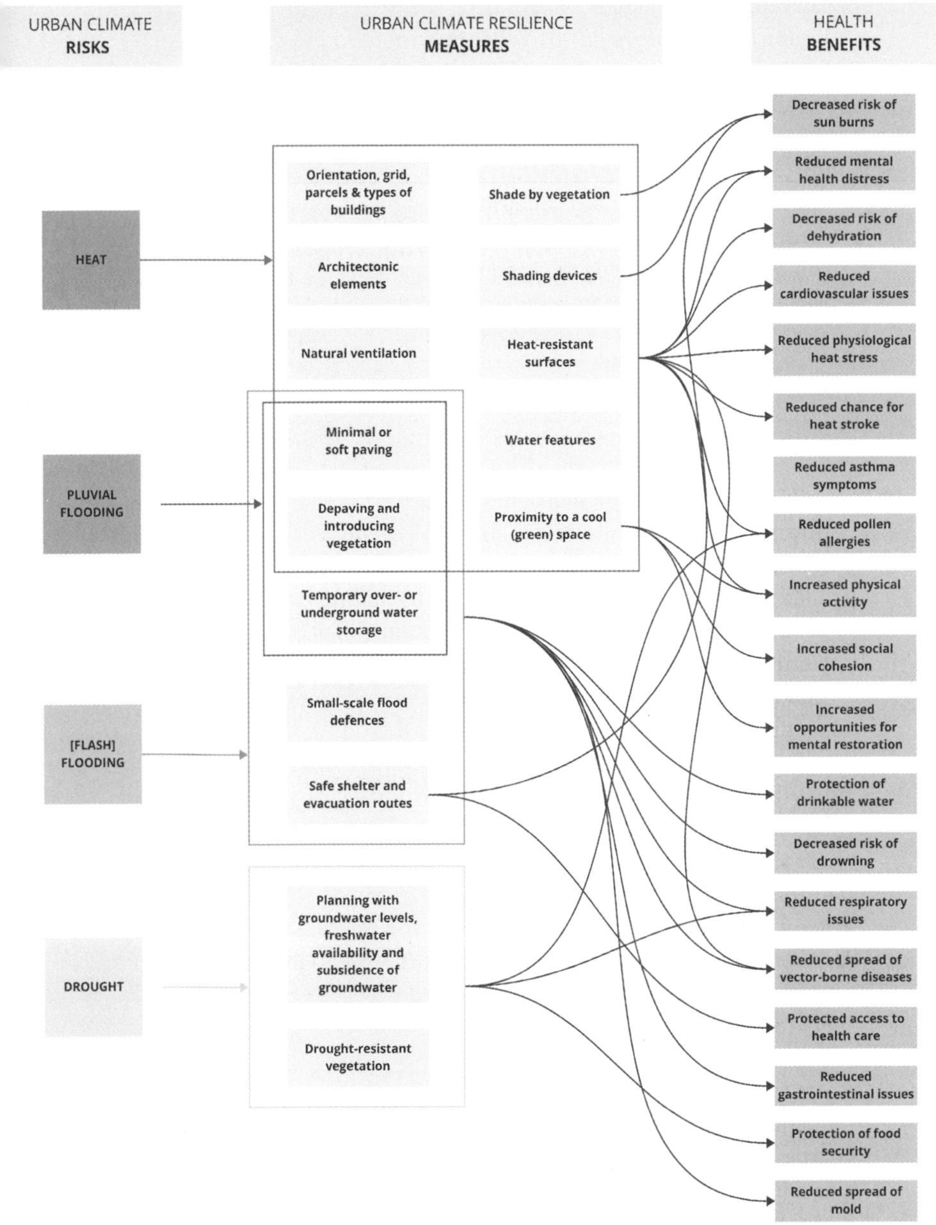

Figure 5: Overview of the main intersections of urban climate resilience and public health. João Cortesão and Agnès Patuano, 2024.

References

Agache, I., Sampath, V., Aguilera, J., Akdis, C. A., Akdis, M., Barry, M., Bouagnon, A., Chinthrajah, S., Collins, W., Dulitzki, C., Erny, B, Gomez, J., Goshua, A., Jutel, M., Kizer, K.W., Kline, O., LaBeaud, A. D., Pali-Schöll, I., Perrett, K. P., Peters, R. L., Plaza, M. P., Prunicki, M., Sack, T., Salas, R. N., Sindher, S. B., Sokolow, S. H., Thiel, C., Veidis, E., Delmoro Wray, B., Traidl-Hoffmann, C., Witt, C., Nadeau, K. C. (2022). 'Climate Change and Global Health: A Call to More Research and More Action', *Allergy, 77*(5), 1389-1407. doi: 10.1111/all.15229

Ahern, M., Kovats, R. S., Wilkinson, P., Few, R., & Matthies, F. (2005). 'Global Health Impacts of Floods: Epidemiologic Evidence', *Epidemiologic Reviews, 27*(1), 36-46.

Barton, H. (2005). 'A Health Map for Urban Planners', *Built Environment, 31*(4), 339-355.

Brandsma, S., Lenzholzer, S., Carsjens, G. J., Brown, R. D., & Tavares, S. (2024). 'Implementation of Urban Climate-Responsive Design Strategies: An International Overview', *Journal of Urban Design, 29*(5), 598-623. doi: 10.1080/13574809.2024.2314760

Calthorpe, P. (2011). *Urbanism in the Age of Climate Change* (Washinton, DC: Island Press).

Cameron, D. & Jones, I.G. (1983). 'John Snow, the Broad Street Pump and Modern Epidemiology', *International Journal of Epidemiology, 12*(4), 393-396. doi: ezproxy.library.wur.nl/10.1093/ije/12.4.393

Corburn, J. (2004). 'Confronting the Challenges in Reconnecting Urban Planning and Public Health', *American Journal of Public Health, 94*, 541-546. https://doi-org.ezproxy.library.wur.nl/10.2105/AJPH.94.4.541

Cortesão, J., & Copeland, S. (2021). 'Urban Climate Resilience', in R. Brears (ed.), *The Palgrave Encyclopedia of Urban and Regional Futures*, 1962-1967 (London: Palgrave Macmillan).

Cortesão, J., & Forgaci, C. (2023). 'Applied Urban Resilience Framework: Operationalizing Resilience in Urbanized Landscapes through Spatial Design', in K. Bishop & L. Corkery (eds.), *Routledge Handbook of Urban Landscape Research*, 195-207 (London: Routledge).

Couch, C., Sykes, O., & Börstinghaus, W. (2011). 'Thirty Years of Urban Regeneration in Britain, Germany and France: The Importance of Context and Path Dependency', *Progress in Planning, 75*, 1-52. doi: 10.1016/j.progress.2010.12.001

Damialis, A., Traidl-Hoffmann, C., & Treudler, R. (2019). Climate Change and Pollen Allergies', in M. R. Marselle, J. Stadler, H. Korn, K. N. Irvine & A. Bonn (eds.), *Biodiversity and Health in the Face of Climate Change*, 47-66 (Cham: Springer).

DeSIRE. (2020). *Mission statement.* Available: https://www.4tu.nl/resilience/research/desire-mission-statement/ (accessed 18 June 2025).

Hiemstra, J. A. (2018). *Groen in de stad: soortentabel* (Wageningen: Wageningen University & Research).

Jacobs, C., Klok, L., Bruse, M., Cortesão, J., Lenzholzer, S., & Kluck, J. (2020). 'Are Urban Water Bodies Really Cooling?', *Urban Climate, 32*. doi: 10.1016/j.uclim.2020.100607

Jenks, M., Burton, E. Jenks, M. & Williams, K. (eds., 1996). *The Compact City – A Sustainable Urban Form?* (Oxford: E & FN Spon).

Jones, B. (2002). 'Capabilities and Limitations of Thermal Models', *Energy and Buildings 34*(6), 653-659 http://dx.doi.org/10.1016/S0378-7788(02)00016-6

Kleerekoper, L. (2016). *Urban Climate Design Improving Thermal Comfort in Dutch Neighbourhoods* (PhD dissertation, TU Delft).

Kluck, J., Kleerekoper, L., Klok, L., Solcerova, A., Loeve, R., Erwin, S., Liu, C., Welter, N., Lopes, M., & Rajaei, S. (2020). *De hittebestendige stad: Coolkit: toolkit voor ontwerpers van de buitenruimte* (Amsterdam: Hogeschool van Amsterdam).

Konijnendijk, C. C. (2023). 'Evidence-Based Guidelines for Greener, Healthier, More Resilient Neighbourhoods: Introducing the 3-30-300 Rule', *Journal of Forestry Research, 34*(3), 821-830. doi: 10.1007/s11676-022-01523-z

Lenzholzer, S., & Nikolopoulou, M. (2020). 'Foreword to the Special Issue on Subjective Approaches to Thermal Perception', *International Journal of Biometeorology, 64*(2), 167-171). doi: 10.1007/s00484-019-01857-0

Luyten, A., Winkler, M. S., Ammann, P., & Dietler, D. (2023). 'Health Impact Studies of Climate Change Adaptation and Mitigation Measures. A Scoping Review', *The Journal of Climate Change and Health, 9*, 100186. doi: 10.1016/j.joclim.2022.100186

Mitchell, R., & Popham, F. (2008). 'Effect of Exposure to Natural Environment on Health Inequalities: An Observational Population Study', *The Lancet*, *372*(9650), 1655-1660.

Mouratidis, K. (2021). 'Urban Planning and Quality of Life: A Review of Pathways Linking the Built Environment to Subjective Well-Being', *Cities, 115*, 103229. https://doi.org/10.1016/j.cities.2021.103229

Mulder, K., Lubelli, B., & Dijkhuis, E. (2023). 'Factors Favouring Vegetation in Quay Masonry Walls: A Pilot Field Study', *Building and Environment, 233*, 110090. Doi: 10.1016/j.buildenv.2023.110090

Natuur & Milieu. (2023). *Bomen voor biodiversiteit. Onderzoek naar bomen in Nederlandse gemeenten*. Available: https://natuurenmilieu.nl/app/uploads/Bomenonderzoek-Nederlandse-gemeenten-Natuur-Milieu.pdf (accessed 18 June 2025).

Nikolopoulou, M., Baker, N., & Steemers, K. (2001). 'Thermal Comfort in Outdoor Urban Spaces: Understanding the Human Parameter', *Solar Energy, 70*(3), 227-235. doi: 10.1016/S0038-092X(00)00093-1

Paterson, D. L., Wright, H., & Harris, P. N. (2018). 'Health Risks of Flood Disasters', *Clinical Infectious Diseases, 67*(9), 1450-1454. doi: 10.1093/cid/ciy227

Schmidt, C. W. (2015). 'Delta Subsidence: An Imminent Threat to Coastal Populations', *Environmental Health Perspectives, 123*(8), A204-A209. doi: 10.1289/ehp.123-A204

Sweco. (2022). *Voorkomen is beter dan genezen: ruim 8 patiënten minder voor elke hectare extra groen*. Available: https://www.sweco.nl/actueel/nieuws/meer-groen-resulteert-in-minstens-62-000-minder-patienten/ (accessed 18 June 2025).

Sweco. (2024). *Een praktische verkenning naar een Nederlandse groennorm. Een vergelijking van tien bestaande groennormen en financiële doorrekening van de drie best beoordeelde normen*. Available: https://natuurenmilieu.nl/app/uploads/Praktische-verkenning-naar-een-Nederlandse-Groennorm-Sweco-2024.pdf (accessed 18 June 2025).

Tamminga, K., Cortesão, J., & Bakx, M. (2020). 'Convivial Greenstreets: A Concept for Climate-Responsive Urban Design', *Sustainability, 12*(9), 2-23. doi: 10.3390/su12093790

Tseliou, A., Tsiros, I. X., Lykoudis, S., & Nikolopoulou, M. (2010). 'An Evaluation of Three Biometeorological Indices for Human Thermal Comfort in Urban Outdoor Areas under Real Climatic Conditions', *Building and Environment, 45*(5), 1346-1352. Doi: 10.1016/j.buildenv.2009.11.009

Wolch, J.R., Byrne, J. & Newell, J.P. (2014). 'Urban Green Space, Public Health, and Environmental Justice: The Challenge of Making Cities 'Just Green Enough',' *Landscape and Urban Planning, 125*, 234-244. doi: 10.1016/j.landurbplan.2014.01.017

World Health Organization (1948). *Summary Reports on Proceedings Minutes and Final Acts of the International Health Conference held in New York from 19 June to 22 July 1946.* Available: https://apps.who.int/iris/handle/10665/85573 (accessed: June 2024).

World Health Organization (2024). *One Health.* Available: https://www.who.int/health-topics/one-health#tab=tab_1 (accessed: March 2024).

Eco-Empathy and The New Delta Work

Joost Adriaanse and Ester van de Wiel

The emergence of a new delta culture and the formulation of eco-social delta futures require 'empathic field work'. Ongoing involvement by the authors with a wide variety of biotopes, sociocultural practices, materials, organisms and 'skilled practitioners' through design research has helped articulate a multitude of eco-social practices for the delta and its human and non-human inhabitants. These speculative practices, framed as 'the New Delta Work', are a product of a collaborative design research strategy called 'empathic field work'. They explore our shared future in the dynamic zone known as the Dutch delta, the collective estuaries of the rivers Rhine, Meuse and Scheldt, heavily influenced by man for safety against floods. Based on these research experiences, the authors call for what they see as one of the most important and highly needed cultural changes of our time: eco-empathy. This sensitivity is cultivated through action. Once developed sufficiently, it will allow us to envision and enact the delta through the lens of care and cohabitation.

Towards a Delta Manifesto: Eco-Empathy and The New Delta Work

The search for the New Delta Work is an ongoing design study that we have been working on over the past three years. It is based on empathic field work, a sensitive joining of the many delta situations that produce both situated knowledge and new relationships. It is a highly productive endeavour that has led to detailed scenarios on participatory coastal regeneration and natural defences, new roles and product lines, as well as co-production models and attitudes co-developed with the many human and non-human Delta Workers we've come to know during these years. We refer the interested reader to our website www.zee-plaats-werk-land.nl, an online platform and research tool, for more details on its progress. Its impact cannot easily be described in words alone, as its most profound result is its contribution to a different way of thinking and acting in the delta. It has engendered a certain growing sensitivity in those who are bound up in it. Its effects are thus implicit and tacit. Nonetheless, describing these largely implicit results is useful. It gives us the opportunity to reflect on certain strategies and allows us and the reader to begin to see and understand the delta through different frames. Words are powerful when put into action, and it is our hope that our words here inspire action in the reader.

In our ongoing search for the New Delta Work, language plays an important role in formulating and exploring the possible futures that this new type of work might lead to. Words, when explored and performed hands-on and in situ, open up many possible worlds. They are kernels from which whole worlds might sprout. In our design research, words inspire action as they describe roles and name tools, products or strategies that get enacted. It is these actions that bridge the gap between the conceptual and the practical and propel the research forward. They generate new knowledge, and play into our existing, tacit knowledge. When put into action, words entangle with the complex delta realities of the present. We might say that present reality and possible future productively meet. Based on our design research actions, we conclude that a manifesto for dynamism is due – one which leads to further action and helps the New Delta Work develop. To learn how to relate differently to the delta, we must engage with it. What we need is hands-on thinking and acting.

Our call is based on our own hands-on and in situ design research experiences in the delta. Figures 1-4 show an example of such an in situ and hands-on exploration: the delta scenario 'Eating against sand hunger'. This involved an eat-work session where taking care of the delta is just a dinner away. A biotopian-chef serves locally harvested shellfish in what can only be described as an 'eat-build menu'. The empty shells of the seafood used in the dinner form the bio-based building material for the on-site creation of anti-erosion reefs on the mudflats. We eat, work and build reefs that form habitats for the many bio-builders in the intertidal zone, who protect the saltmarsh from eroding and in turn strengthen flood safety. New words, roles and strategies are put into action in this 'sketching-in-dialogue' session to further explore participatory coastal strategies and forms of the New Delta Work. With contemporary Delta Workers, we physically explore the Delta Work of a possible future. The participants included an offshore industry CEO, a geographer and aspiring island builder, the daughter of drs. A. C. Drinkwaard – a prominent delta figure and biologist known for his banned speech 'Geeft ons geen Oosterscheldedam, maar wat dan?' ('Don't give us an Eastern Scheldt Dam, but what then?'; Drinkwaard, 1965) – two bio-based architects, a marine ecologist and two design researchers. They take on the roles of biotopian-chef, reef building recreationist, patchiness ecologist and habitat workers.

All of them are 'skilled practitioners' as described by Rietveld and Kiverstein (2014), with their own sociocultural practices, experience and skills. When skilled practitioners from different fields of expertise and sociocultural practices come together on site in a specific material setting, things start to happen. It can lead to the joint discovery of new possibilities for action for and by Delta Workers. This scenario explores how known nature-based coastal strategies in the delta (reefs as coastal defence components) might interweave with the various ways of life in the delta. In other words, this scenario explores how nature-based solutions (European Commission, 2024) might lead to a nature-based culture and future. As such, it leaves behind the solution frame and focuses on the possibility of forming and changing ongoing delta relationships. In a collective effort of eating and building habitats for the non-human bio builders (flat oysters, Japanese oysters, mussels), a dialogue with and about the delta is sparked. A hands-on exploration unfolds. We are where all delta worlds seem to meet: the intertidal flats outside of Yerseke.

Offshore industry CEO and biotopian-chef Jaap intertwines his professional offshore knowledge with his culinary and personal experiences as he prepares and serves locally harvested oysters on the grounds of machine factory Bakker in the fishing port of Yerseke. Patchiness ecologist Jim, a marine ecologist, chimes in and elaborates on the local sediment erosion reality and his search for new nature-based spatial arrangements that might be successful in protecting the estuarine nature area just a few kilometres east. In a back and forth and in multiple small conversations, design researchers, geologists and architects explore how our role as reef building recreationists might form a super-local participatory coastal practice. Many hands make light work and salt marsh, mud flat and reef are better suited for manual labour than heavy machinery. What might an extensive mussel reef fishery look like when combined with ecological monitoring and reef recreation? What new tools do we need to facilitate that? What type of landscape might be the result? As we eat and talk, the tide has gone out. We move towards the edge of the mudflat to position the fruit of our recreational labour: a reef-sock knitted from locally grown flax, filled with the empty shells of our meal.

As they secure the reef-sock in place, patchiness ecologist Jim and reef recreationist Idco, a geographer, discuss the possibilities of reef building recreation as an intergenerational form of both reef and island making. Many people living in the delta have no intimate knowledge of the relations between various biotopes, ecosystems and their well-being, let alone their entanglement with our current ways of life, spatial planning and delta activities. Long-lasting collective and participatory scenarios such as these might be able to begin to change that, moving coastal defence, nature conservation, restoration and spatial development out of the closed realms of technical expertise and into the shared public domain. Changes in our living environment are inextricably linked to changes in living with our environment. To create a shared future where the many different human and more-than-human ways of life are attuned to one another, we must find ways to entangle ourselves and engage in delta relationships.

Entangling as Design Work

The delta is continuously being enacted. Its performers range from tidal flows, marine bio builders, migratory birds and salt marshes to oyster fishers, dredging ships and all sorts of organizations, institutions, traditions, humans and non-humans that are bound up in them. Our ongoing action research is performed together with these Delta Workers, both human and non-human. In a reflective hands-on conversation with them we explore how each Delta Worker contributes to shaping the delta. Through their words, frames, protocols, movements, actions and materials, and through their stories, relationships, interactions and the many ripples they create in the dynamic watery world that is the delta, we get acquainted with it. We have successfully used this continually evolving strategy in previous projects (Verstraete et al., 2021; Hamers et al., 2021). In this case it is applied as an open-ended exploration of the possibilities of 'being delta' in its many forms. Our goal is to learn how the delta 'is' and how it could be, to propose possible futures that are strongly connected to the present and are composed of the ways of life of the many Delta Workers that we have come to know.

This process of entangling ourselves with the complex practices and delta lives of these Delta Workers is in part reflective. As they move through biotopes and work fields, operate and think in varying scales and live with different time horizons, engaging with them gives insight into how things are being done. It is a way of experiencing the myriad flows and goings-on that enact the delta as we know it. It is reflection in and on action. As such, it allows for a reconceptualization of the delta and the creation of alternative pathways through design action. This is where the call for dynamism rings loudest. We strongly encourage the human Delta Workers reading this publication to engage in what we call 'empathic field work' (Figures 5-7). It requires us to work alongside other skilled practitioners in the field and learn about their ways of working and seeing, their vocabulary, their way of framing the delta and their contribution to its many forms by quite humbly 'being there' and working alongside them with our senses attuned.

As such, we've not only experienced the world of the mussel fisherman or the wild seaweed cutter, but also the lives and entanglements of the mussels, oysters and seaweeds themselves through adopting this attitude. Working alongside these Delta Workers, you experience the local effects of ordinances and permit systems, and see how bristle worms' lives, baiting practices and saltmarsh morphology are intertwined. You feel the strength of the tide and experience the beautiful colours of seaweed. You get a taste of the practical reality of working on a mudflat, of navigating tidal flows and the many other entanglements of systems, people, places, things and forces hands-on. Our proposals and scenarios that explore the New Delta Works are all based on empathic field work. Through it, the swarm of involved Delta Workers grows, allowing for hands-on explorations like the one in Figures 5-7. Empathic field work allows us to move between realities and find frames and practices to bisociate and formulate new possible worlds.

Bisociation, a term coined by Koestler (1964, 35), means to productively cross two frames through which one can perceive a situation – 'two self-consistent but habitually incompatible frames of reference' – out of which new knowledge, insights and possibilities might emerge. Bisociation as a design strategy is discussed in more detail in our publication on caring confrontations (Hamers et al., 2021). For now, it suffices to say that empathic field work and bisociation rely on the ability to be open, interested, unbiased and reflective. Our role has been that of a conduit, finding productive overlap, friction or possibilities between different fields of experience and expertise, in order to inspire and invigorate one another in search of possible futures. But it is not just design researchers that ought to engage in empathic field work. This is in part the essence of our call.

Life in Interrelation

We need to learn to 'live with'. We need to learn to join in and live in interrelation. In a delta that has been most dominantly shaped and altered by human actions firmly rooted in engineering traditions and safety frames, we need to reorient ourselves and experience the delta from multiple perspectives and through different frames. The common ways of framing the issues of concern in the delta and our responses to them have more often than not been through a lens of techno-optimism. Yet the current state of the delta calls for a new attitude towards things. We need to cultivate eco-empathy and learn to see the delta for the ongoing and dynamic complexity that it is: a whirling and vibrant assemblage of things, matter and all kinds of human and non-human actants (Bennett, 2010). Ever moving, always in process. However, thinking in processes, continual change, stochasticity, fluidity and dynamism has proven hard for us humans and our institutions. Hard borders have become a cultural reflex. Abstraction, reductionist approaches, rigid frameworks and simplification are ingrained in our ways of thinking about and acting in and with the delta. Through time we have come to expect and demand a delta that obeys our rigid frameworks.

Through our collective and open-ended exploration of the ways of the delta and the act of proposing possible futures in close dialogue with those that live and work in the delta, we make way for a new dynamism: an intimate entanglement with the world, a shift from techno-optimism to eco-empathy. Eco-social thinking and doing calls into question existing categories, frames and attitudes. Yet, the rigid border persists in our thinking. All too often we place ourselves outside of nature, outside of the living world. While we are inextricably linked to and form part of the biotopes and habitats of others and vice versa, we remain emotionally disconnected from this dynamic reality. The search for the New Delta Work is a search for an eco-social future and a call for eco-empathy. This call for eco-empathy is a hopeful call, for it is in our nature. For us to reconnect with the one world we all inhabit and to understand its intricate entanglements, we must become active participants in its becoming and learn to see, feel and be in interrelation.

Proposed possible futures and scenarios through which we can learn to live in interrelation must have a footing in present reality. No far-flung and disconnected futures that feel so far away they become impossible to imagine reaching, but scenarios that can emerge from the present, right here, right now. Finding room for the future within the present leads to scenarios that are rooted in the critical space between what is and what could be. Such scenarios speak to those whose world you collaboratively propose to see differently. Ecosystem service duty (Figures 8 and 9) represents one such scenario. It speculates on governmental protocols and addresses contemporary civil servants in the delta. Ecosystem services are often seen from an egocentric perspective, as the services that ecosystems provide to humans. Yet we are an intricate part of these ecosystems. This scenario challenges the bureaucratic order of Excel sheets, the permit apparatus, the national governments, and calls into being the Agency for Eco-Social Affairs. You have been called to Ecosystem Service Duty!

This speculative vision is conceptualized as a workstation for the civic servant, calling up civilians for ecosystem service duty. This call, by means of a formal letter, goes out to all eligible inhabitants of the country, drafting them for a periodical contribution to the various ecosystems within the Netherlands. The workstation exhibited at the International Architecture Biennale Rotterdam 2024 contains the letter, notifications, videos, models, shells, examples of habitat work and workwear. After the exhibition it will find its home in the offices of the Province of Zeeland, alongside contemporary civil servants' workstations.

Meanwhile, the exploration of this possible future continues as eco-system-service-duty t-shirts dyed with local seaweeds are given to Delta Workers, along with a personalized call to perform their ecosystem service duty right now, in the present. These Delta Workers range from those within government and politics to those working in the actual water of the delta. For the Delta Workers closely involved in the articulation of this scenario (through empathic field work, tagging along in their day-to-day delta practices) these ecosystem service duties are often a reframing of activities they already perform. For instance, the work of marine ecologists like Jim is often already geared towards gaining insight into the wellbeing of ecosystems and their role in coastal morphology (van Belzen, 2023). However, his call to eco-system-service-duty focuses on exploring the possibilities of opening up his scientific fieldwork for the public to participate in. This makes him one of the Delta Workers with whom others could possibly perform their ecosystem service duty, thus contributing to the saltmarsh biotope and strengthening human-saltmarsh relations.

A New Dynamism

Scenarios like those described in the figures shown here inspire a sense of possibility. They point us in possible directions for answering the questions of how to collectively become comfortable with dynamism and learn to live with the dynamic natural world. So how do we cultivate eco-empathy? How do we perform it and put it into practice in each of our own lives? If any place can teach us dynamism, it is the delta with its intertidal zones. Twice a day land becomes water and becomes land again, flocks of birds migrate in and out, shellfish and worms surface and dig back into the mud, while seaweeds sway with the tides. In this zone, the delta's dynamism synchronizes with our own sense of time. The longer you spend there, the more you start to see and feel. It is time to become active participants in the living delta. It is time to experience the human and non-human entanglement that is the delta. It is time to become delta.

The Delta Manifesto is being 'written' as we speak. At this moment it consists of a table of contents, a series of provisional chapter titles (each of which will spawn new projects), multiple email conversations, images, drawings, future scenarios, a lexicon, job and role descriptions, a new speculative branch of government, a product line, the remnants and shells of the flat oysters that have provided the material for recent reef restoration projects, clothing, models, various in situ actions and a set of working principles. More importantly, the manifesto is an attitude that grows in the minds and bodies of the many Delta Workers that continue to be involved in this project. Parts of the manifesto have been given, loaned or shipped to other Delta Workers throughout the delta, in line with the idea of creative entanglement. The manifesto is a contribution to what we see as one of the most important and crucial cultural changes of our time: eco-empathy. This sensitivity is cultivated through experiencing, seeing, feeling, hearing, engaging with and relating to the life around us, both human and non-human. Eco-empathy can grow through 'correspondence': dynamic and responsive engagement with the world and its many unfolding ways of life (Ingold, 2011). When grown sufficiently, it allows us to no longer see the world through the frames of domination, but through those of care and cohabitation.

Developing this sensitivity is an experiential skill, so to speak. Therefore, we follow a set of working principles and attitudes that aid us in this process. Inspired by the investigative practices and philosophies of Bruno Latour (2004; Latour & Weibel, 2020), Henk Oosterling (2020), Isabelle Stengers (2018) and Tim Ingold (2010; 2011), we adhere to several guidelines for hands-on thinking and doing. We encourage you all to actively apply them in your own field of work. For now, while we continue working towards a manifesto, we leave you with the following important working principles that are part of the ever-growing delta manifesto.

/ Embrace complexity, do not simplify
/ Embrace what we don't know
/ Be curious
/ Feel time (time, timing, rhythm, slowness)
/ Be open and not biased
/ Be generous; share

/ It is never finished (always in the making)
/ Compose and bring together
/ Move between scales and disciplines
/ Collaborate
/ Go out into the world!

The national Delta Commissioner, Co Verdaas, emphasized (albeit implicitly) the importance of such working principles and explorations in his keynote lecture during the Southwestern Delta conference on 4 July 2024, when he proposed to be adaptive, to not strive for perfection, to just start, try out and learn. He summarized these principles with the phrases 'stumble intelligently' and 'find the fuss', echoing the words of both Bruno Latour's matters of concern (2004) and Donna Haraway's call (2016) for staying with the trouble. Our call to learn to think, act and live with dynamism seems to have a common goal with the words of the Delta Commissioner: an exploration of new narratives and the start of a new discourse. As far as we're concerned, these narratives draw not from the frameworks of domination that have inspired much of the engineering in the delta over the last centuries, but from a sense of dynamic entanglement and eco-empathy. As such, and when activated through in situ, collaborative actions, and materialized in speculative prototypes, tools and scenarios, they contribute to the composition of a dynamic and eco-social delta culture.

In concluding this call for eco-empathy, the Dutch word 'meebewegen' comes to mind. 'Meebewegen' can be translated as 'to adapt', 'to move along with' or 'move along together'. It conveys the ideas of adapting to changes, being flexible, moving in concord with and dynamically corresponding with. In it lies the call for a reorientation of our relationship with the natural world. With it, it perhaps inspires a reinterpretation of the Deltares research track *Meebewegen* (Deltares, 2020). As Ingold (2021, 22) writes: 'we have forgotten how to correspond with the beings and things of which [the natural world] is comprised'. In our preoccupation with 'the interaction between ourselves and others . . . we have failed to notice how both we and they go along together in the current of time'. As such, 'meebewegen' in the context of the delta implies to co-exist, to co-inhabit and to embrace more-than-human perspectives in an attempt to learn to eco-socially participate in the work of the swarm of bio-builders and Delta Workers that enact the delta. Towards eco-social delta futures!

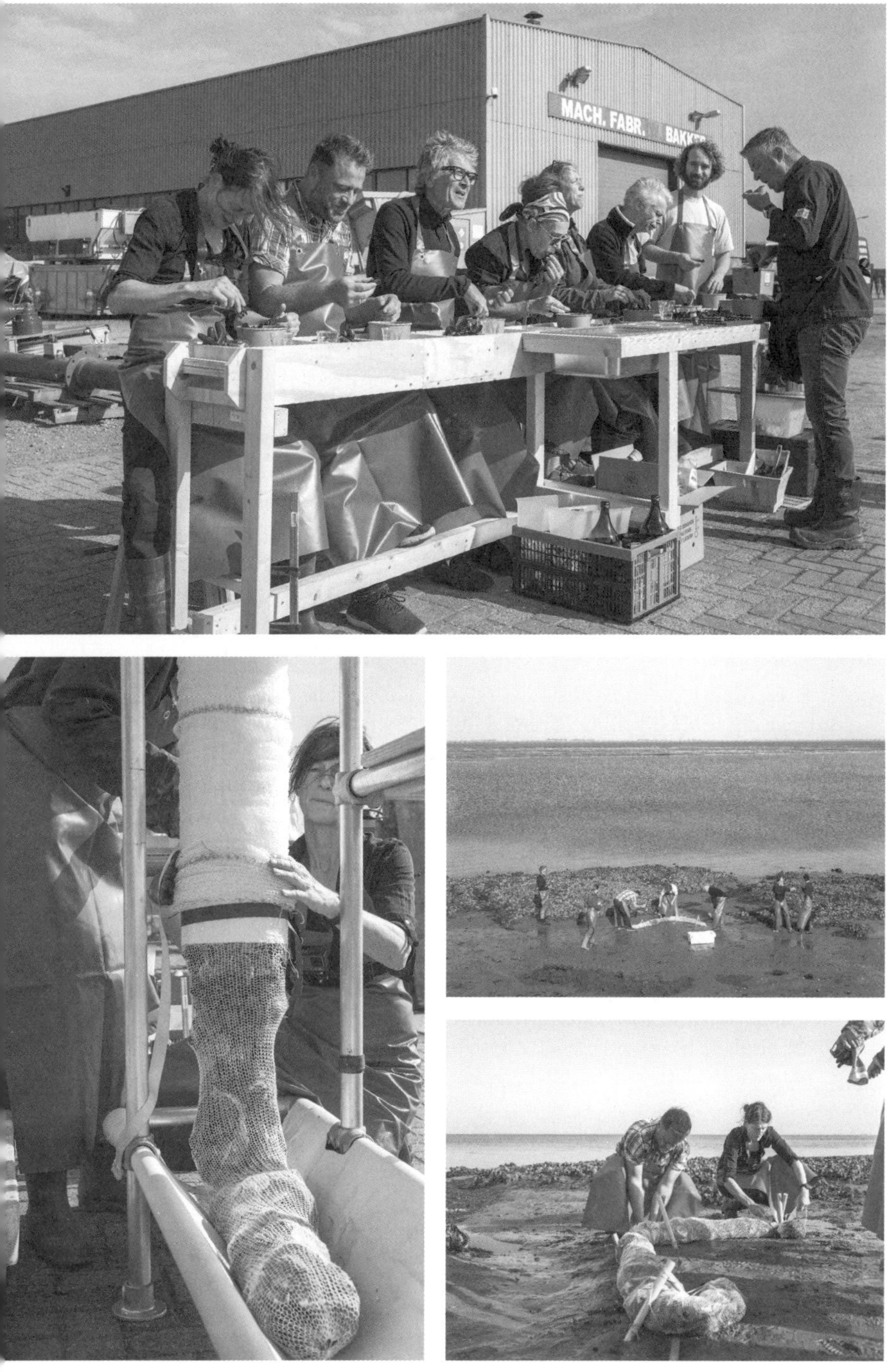

Figures 1, 2, 3 and 4: A 'sketching-in-dialogue' session to explore the delta scenario 'Eating against sand hunger' and its participatory coastal strategies as form of the New Delta Work. Yerseke, 2023. Photo: Walter Herfst.

Figures 5, 6 and 7: Learning alongside the human and non-human Delta Workers through empathic field work. Dutch Delta, 2023. Photo: Joost Adriaanse (above) and Ester van de Wiel (below).

Figures 8 and 9: Drawings, design and materialization of the delta scenario 'eco-system-service-duty' during the International Architecture Biennale Rotterdam 2024, titled 'Nature of Hope'. Rotterdam, 2024. Photo: Jacqueline Fuijkschot.

Figure 10: Seaweed farmer Joost. Veerse Meer, 2023. Photo: Suzette Bousema.

Figure 11: Impression of the scenarios that are currently being envisioned. Studio Ester van de Wiel + Studio Joost Adriaanse, 2024.

Video showing the ecosystem service duty letter being written. Vimeo, Deltawerker. Studio Ester van de Wiel + Studio Joost Adriaanse, 2024.

Videos about the development of the delta scenario: Eating against sand hunger. Vimeo, Deltawerker. Studio Ester van de Wiel + Studio Joost Adriaanse, 2024.

References

Bennett, J. (2010). *Vibrant Matter: A Political Ecology of Things* (Durham, NC: Duke University Press).

Deltares. (2020). *Adaptatie aan zeespiegelstijging: Meebewegen.* Available: https://publicwiki.deltares.nl/display/KWI/Meebewegen (accessed 17 June 2025).

Drinkwaard, A.C. (1965). *Geeft ons geen Oosterscheldedam, maar wat dan?* [unpublished lecture]. Available: https://sharedconcepts.nl/wp-content/uploads/2023/10/Toespraak-drs.-A.C.-Drinkwaard-22-26-januari-1965---origineel.pdf (accessed 17 June 2025).

European Commission (2024). *Nature-Based Solutions*. Available: https://research-and-innovation.ec.europa.eu/research-area/environment/nature-based-solutions_en (accessed 17 June 2025).

Haraway, D. J. (2016). *Staying with the Trouble: Making Kin in the Chthulucene* (Durham, NC: Duke University Press).

Hamers, D., van de Wiel, E., Adriaanse, J., & Verstraete, G. (2021). 'A Caring Confrontation: Re-Ordering as a Design Research Strategy Toward a Circular City', *GeoHumanities*, *8*(1), 250-264. https://doi.org/10.1080/2373566X.2021.1915700

Ingold, T. (2010). *Bringing Things to Life: Creative Entanglements in a World of Materials*. NCRM Working Paper Series, no. 15 (Swindon: Economic and Social Research Council). Available: https://eprints.ncrm.ac.uk/id/eprint/1306/1/0510_creative_entanglements.pdf (accessed 17 June 2025).

Ingold, T. (2011). *Being Alive: Essays on Movement, Knowledge and Description* (London: Routledge).

Ingold, T. (2021). *Correspondences* (Cambridge: Polity Press).

Koestler, A. (1964). *The Act of Creation* (New York: Penguin Books).

Latour, B. (2004). 'Why Has Critique Run out of Steam? From Matters of Fact to Matters of Concern', *Critical Inquiry, 30*(2), 225-248. https://doi.org/10.1086/421123.

Latour, B., & Weibel, P. (2020). *Critical Zones: The Science and Politics of Landing on Earth* (Cambridge, MA: MIT Press).

Oosterling, H. (2020). *Verzet in ecopanische tijden: van ego-emancipatie naar eco-emancipatie* (Amsterdam: Lontano).

Rietveld, E., & Kiverstein, J. (2014). 'A Rich Landscape of Affordances', *Ecological Psychology*, *26*(4), 325-352. https://doi.org/10.1080/10407413.2014.958035

Stengers, I. (2018). *Another Science Is Possible: A Manifesto for Slow Science* (transl. S. Muecke). (Cambridge: Polity Press).

Van Belzen, J. (2023). *Struggle & Emergence: On Critical Dynamics and Geometry in Tidal Ecosystems* (PhD Thesis, Utrecht University).

Verstraete, G., Adriaanse, J., van de Wiel, E., & Hamers, D. (2021). 'Assembling Researchers in Design and the Humanities in a Circular Ecology', *GeoHumanities, 8*(1), 232-249. https://doi.org/10.1080/2373566X.2021.1915699

. . . En de zee was niet meer, Bert Haanstra, 1955

Rommy Albers

. . . En de zee was niet meer (*. . . And the Sea Was No More*) is a kind of counterpoint between *Dijkbouw* and *Delta Phase 1*, the two other commissioned films Haanstra made about the Netherlands and its struggle against water. Both the latter films show the construction of hydraulic works: the traditional dike construction in the former Zuiderzee and the large-scale caisson dams rising as part of the Delta Works in Zeeland and Zuid-Holland. These hydraulic works were part of the effort to protect the Netherlands and at the same time gain more farmland, after the storm disasters in 1916 and 1953 had shown the need for defence and the intensification and mechanization of agriculture demanded a new arrangement of the land. The latter requirement was met with the closure of the Zuiderzee and the reclamation of the IJsselmeer, long before the Delta Works had even begun.

But this development also had a downside. Many port towns on the former Zuiderzee were cut off from the outer sea. Fishing fleets had to move, old ports silted up; an old culture was in danger of disappearing. There were protests against this from the very beginning and we see the repercussions in film. In pre-war films like *Dood water* (Gerard Rutten, 1934) and *De laatste dagen van een eiland* (Ernst Winar, 1938), but also in post-war films like ... *En de zee was niet meer*, the loss was present.

... *En de zee was niet meer* was supposed to be about the old crafts and customs in the towns and villages along the IJsselmeer. Haanstra sticks to his task and goes further. In the film, he follows the course of man's life: birth, childhood, adulthood, old age – and death. Then, he makes a hard transition. He cuts from the earth thrown on the coffin to a gripper dumping the last cargo of clay into the last hole in a sea dike. The old times of fishing, vivacity and prosperity are over; a new, uncertain time is dawning.

Dialogue with Digna Sinke

Digna Sinke and Sophia Arbara

This dialogue took place in June 2024 at the Faculty of Architecture and the Built Environment at TU Delft, following the Reporting the Delta screening series held during the 2023/2024 academic year. The series featured selected films about the Dutch delta, accompanied by discussions with invited guests. One of the featured screenings included Digna Sinke's film *Weemoed & Wildernis* (*Wistful Wilderness*; 2010), which served as a starting point for this post-event conversation between Digna Sinke and Sophia Arbara. The dialogue reflects on the role of filmmaking as a mode of inquiry, the preservation of memory through landscape and film, and the layered spatial and cultural transformations within the Dutch delta.

Digna as a Filmmaker

Sophia Arbara: On your website I read: 'I'm a filmmaker. I see that broadly. I don't know where it ends and where it begins . . .' What does being a filmmaker mean to you?

> Digna Sinke: For me, filmmaking is a way of life. It is about observing things, human beings and the world around me. Exploring while asking myself questions about it. What is it? Can I be part of it, or do I reject it? Do I feel connected to it or not, and why? In a way filmmaking is not only a profession, it's a way of life. Nowadays, filmmaking is very professionalized. When I went to the Dutch film school, the *Nederlandse Filmacademie* in Amsterdam, you could pick between tracks A and B. A was more about designing, production, writing and directing; B was more on the technical side. I went there because I was a kind of a fanatical photographer. I didn't know much about it, but I very much liked watching and exploring. I liked going out in Utrecht, where I lived, with my camera on weekends and free afternoons, choosing what kind of things I would like to take a picture of. It started there. Or perhaps it

> started even earlier, when I was drawing quite a lot as a kid. I was not drawing the usual stuff for a little girl – no fairytales. Instead, I drew the haystack of our neighbour. Our neighbour was a farmer; he kept hay on a dyke to feed the animals. I liked that shape. Back then I was seven years old, and I didn't realize that was not a 'normal' thing for a little girl to do. Only later I realized it was quite strange.

Is filmmaking also a way for you to describe and connect with your own territory?

> I try to describe my territory, but I also don't know exactly how big it is. Maybe it's endless. It's there, but it's also not. To me, it also means to go over the boundaries.

In the film *Weemoed & Wildernis* (2010), in Tiengemeten, the dikes offer the boundaries.

> Yes, this is also connected to my youth. I was born on the island of Schouwen-Duivenland. Later I moved to the peninsula of Sint Philipsland and then to another peninsula, Walcheren. Later I moved to the mainland, in Utrecht. So, for me, having a territory with boundaries is somehow the normal situation. In Tiengemeten, beyond the dikes there is the wilderness, an imaginary. But imaginaries are in a way everywhere, not only beyond the dikes, and not only spatially.

You mention that filmmaking is part of who you are, navigating between your own personal stories while trying to understand the world, asking very genuine questions. Is understanding the outside world connected, in a way, to understanding who you are?

> Filmmaking started as a way to understand the world around me, but also to feel a sort of connection with it. I always felt kind of like an outsider, and the only way to solve that and find peace in these thoughts is by trying to understand it. To include myself, to be part of the rest of the world. But then you have to work; it doesn't just come to you. You have to come to the world. You have to find out how it works, how things work and what you can do to be part of it. And then, you also have to ask yourself: 'but who am I that I want to do this? Why do some people have questions and other people have no questions at all?' That is strange.

What about taking a stand through filmmaking on issues such as spatial transformations, to raise awareness or as a kind of activism?

I think awareness is the right word. I am not a real activist. Sometimes I am, but not really in my films. What I try to do is to find neglected aspects and indeed create awareness. I aim to show people a wider world than the one they read about in the papers. Film can be more than that, of course. You can make very activist films. And in the 1960s and 70s, I was close to activist film groups making films about, amongst others, housing speculation and urban development. People from Amsterdam had to move to new cities due to all these developments. We recently rewatched the films with a young audience and young people were very surprised that these films were made so long ago and the issues are still present today. We still have to tackle the same issues. At that time, there was the 'Club of Rome' and if you re-read their report, you see a lot of questions that are similar to what we're dealing with now. What has happened in all this time in between?

Was there more optimism then, which is gone today?

Yes, but it was rapidly gone, that optimism. I am not an optimist; sometimes I like to be one, but I can't convince myself.

On Memory and Preservation

Time and transformations over time are very present in your films. Why do you think this concept is important for you?

I have not the slightest idea, I don't know. I was always interested in history. I am from a real Zeeland family. My great-grandmother was a really old woman. I also met her sisters. They were around 90 years old; this was considered very old in the 1950s. They had all kinds of stories about places that I knew because I lived in the same place. It fascinated me that they could tell stories about what I could partly still see. An old barn, a part of a farm, an old tree, part of a road, stones that they used for building the dike. My grandmother always told me a story about my house in Zeeland. She told me that when she was three years old, they made new windows in that house. Now we think they are quite small, but for that time, they were very big. So big that the children from the kindergarten were calling it the 'ice cream shop'. So, I know exactly when it happened. My grandmother was born in 1897 – I know those new big windows were made around 1900. That fascinates me – I always liked it. You see a building and then you hear the stories about it – you create an imaginary of what was once there. Don't forget that this was also a time before television, so people were telling stories.

Storytelling was normal. Not only fairytales, but also weird stories about strange accidents; about anything.

I read that you also like to preserve and keep all sorts of things, both material and immaterial. Is film for you a way to preserve memory?

I posed a similar question to an American guy whose solution is to take photos of everything with his telephone and keep the pictures, but not the physical things. And my final conclusion in the film *Bewaren – of hoe te leven* (*Keeping and Saving or How to Live*; 2018), where my mother is touching a piece of fabric, is that it's also about feeling and touching things. It is about the physical aspect. It is not only about looking, but also about sensing things. This is so different from digital preservation. It is not only about memory – I see preservation as a duty, also in my own family. If you want to know who you are, then you have to know where you come from. So, your own history is important for who you are now and, again, for what your connection to the world is.

Do you think there is also a generational aspect in terms of physical versus digital preservation? Do people today preserve only in the digital sphere?

Yes, it could be, and we don't know how this is going to change in the future. My family told me the stories when there was no television. Today, nobody tells those stories anymore. If someone tells a story, it is more in a scientific way, on rural preservation for example. Not among family, on a birthday party. That's really gone. So maybe even that way of keeping memories is gone.

This way of keeping memories and preserving everyday stories also has an archival research component in it . . .

It would be a pity to destroy it. Do you know the book by Orhan Pamuk (2008), *Museum of Innocence?* It is about collecting things. He wrote the book, a love story; and apart from the book he made a museum. His own museum, similar to what I do with my house. You can still visit it in Istanbul. It is his love story from the book, a semi-autobiographical love story about Pamuk, son of the director of the lemonade factory, who falls in love with a girl. That's the girl in the love story of the book. And then he collects things about that lemonade factory. He collects things about the films he went to see with that girl, in movie theatres in Istanbul that do not exist anymore. He collects cigarette butts because he smoked those cigarettes with his love. The kind of costumes he wore, everything is there in a vitrine. It's amazing. It's kind of fictional, but also nonfictional. Collected items, but in a very personal space.

Perhaps you do the same with your house, which connects to the films?

In a way I do the same, yes. My house is connected to the films. I held an 'open garden' with a little exhibition in the village where I was born, in Zonnemaire. Sometimes I give a kind of lectures about the history of the village, with old folks from my family, because it's so interesting. You see those family photos, let's say the fortieth anniversary of someone's marriage. But behind the group of people, you see the houses, you see the village, you see a dike. There is a kind of history behind the family photos, so I like them very much. It is also about temporality. You can see things now, but maybe tomorrow they are gone. In a way, it's so amazing that we all live in that world, with all those changes we hardly see, but they are always there, at every moment. That is amazing and fascinating, and also a pity, that they will be gone before you have a chance to watch them exactly.

That's also maybe a core part of your films, right?

It is about how short our life is.

In terms of preserving collective memory in the Dutch landscape, what should we preserve in this territory?

I like it very much when you can still see the history of a landscape in the landscape now. Of course, things change, that's the way of it, but I like it very much when I still can think and see – at least partly – what was there long ago. Now in Zeeland we have the UNESCO Global Geopark. There is a lot to see; the man-made landscapes but also old creeks. There is one just behind my garden. If you know it has been there, you can see it. It is not a creek anymore, but a small waterway. It has been there since 1300, so that is 700 years ago. That's interesting, you can see parts of life that existed 700 years ago . . . Maybe you even have to keep some big fields in the Flevopolder for future generations. I think that man-made landscapes are very beautiful places, lots of parts are man-made and they are great. You see a lot of ditches and they are so typically Dutch in certain areas.

They are all man-made, a very nice and interesting part of the landscape.

On Socio-Spatial Transformations in the Delta

In *Weemoed & Wildernis* (2010) we see constructed landscapes, reflecting the Netherlands' largely man-made nature.

Of course, there is almost no area in the Netherlands that is untouched. Even the dunes are touched, even though they are maybe the most 'original' nature. My Canadian family came to the Netherlands and I showed them Tiengemeten. They were so surprised that in such a small country like the Netherlands you have such different landscapes. You've got polder landscapes, dunes, woods, industrial areas, beautiful towns. All in a very small piece of land. A Siberian filmmaker working in the Netherlands once told me that we have much more nature than in Siberia. There, there are only monocultural woods. Here, it might be that you designed it, but it's there, a more diverse landscape.

Growing up in the Dutch delta, how did you perceive the changes, especially after the execution of the Delta Works?

As a child I experienced the great damaging flood: 1900 people died in that flood in 1953. It was not so strange that they had to do something about it – water was everywhere!

So, any response at that time would have been a good response?

Yes, of course. That's how I grew up, that is what everyone told us. And I really grew up seeing those big machines, making new dikes, changing the landscape. I saw the first dam; I think between Walcheren and Noord-Beveland. I was a child and we went there by bike to see what people were doing. We admired all the busy people building new dams to connect two islands. They were heroes in a way; they saved us. I was three years old and our house was full of water. I remember that.

Looking back now at all these transformations, what is your reflection today?

It is interesting that they never made a plan to build a dam in the Western Scheldt. It's impossible because it's too important for the economy of Antwerp. It followed the money, that's how it works. If you want to have a vivid agricultural area, then you have to build a dike and dams, otherwise you lose that land. Building the Eastern Scheldt dam with open barriers was partly also for economic reasons; oysters and mussels were a prosperous industry.

What they did in Tiengemeten was different. Every part of the Netherlands has to have a specific percentage of nature reserve. There was even a law, I think from 1995. I was there from the beginning. Five reports were written by RIZA (Main Department of the Inland Waterways Service), the organization doing the research on how to make a new area, how to ensure ecological quality and infrastructure, but also recreation. They also made a report about the buildings – there were seven farms, but quite old; the initial idea was to demolish them all. What surprised me was the way they have really organized that island very thoroughly. They moved the top layer of the soil about 60 cm because it was too fertile to turn it into nature. If you keep that soil, you can only grow nettles and blackberries, and they wanted to have more diverse nature. They used the soil to build dikes and removed one road, an enormous, long asphalt road – a lot of work. They put the asphalt in a breaker in order to reuse the asphalt, and instead of the road they made a creek. Yes, you can make a creek, but it's bizarre. They had to cut the trees because they would die due to too much water coming in and visitors would not like to see dying trees. So, they cut them all.

With your films, you give a voice to people. Is this intentional; giving a voice to communities or people that may not be heard otherwise?

In a way, my mother is an important part in the film *Bewaren – of hoe te leven* (2018). I really think I gave her the opportunity to say something about a way of living, a way of looking at things which is quite unique. I give people from the past a chance to explain how valuable things were to them. Now we can buy everything everywhere. You can fill in a form on the computer and it comes to you; they deliver it to your house. But in that time, you could be so happy with a small thing because it was special; now you can order whatever you want online. So, yes, I give her a voice, talking about a period that's gone, and I think that's also in the film about industrial archaeology, or for example in Tiengemeten where I give a farmer a voice. You can feel, in the film, how important that island and the work he's doing are.

On Reporting the Delta

Linking your work with the Reporting the Delta project, how do you see the attempt to bring arts and sciences closer together?

> It's difficult for me to judge, because I am just coming in. I like the initiative very much, of course, and I very much like that connection between artists, filmmakers, and science. It's always been there, but it's coming again. Part of the KNAW (Royal Netherlands Academy of Arts and Sciences) is for artists, so they know that there is a kind of connection. Research, exploring things, that's what the connection is between those two groups. They do it in a totally different way, but there they have something in common. I like that.

What do you think they have in common?

> The exploring part; trying to find out, trying to understand, analyzing, curiosity, that's what the connection is. I liked the films that were shown in the Reporting the Delta project. I only knew the 'new' film about Nagele [*Terug naar Nagele* (*Back to Nagele*; van Gasteren, 2012)] and I knew, of course, Louis van Gasteren. I was pleased to see the 'old original' film [*Een nieuw dorp op nieuw land (A new village on new land*; van Gasteren, 1960)]. The discussion that followed the screening was a little bit strange to me, in the sense that there was some criticism on the houses because they were too common. They were very modern in that period and with very new ideas about how you would build up such a community. It was not really judged well, I think, not exactly placed into that period in time and context. But I can imagine that since there were all young people without lots of knowledge about that period, now it seems like nothing special, but it was something special in that time.

It's important to see things also in context.

> Yes, of course, on the one hand, but then you may also see it from today's perspective. I was surprised at the screening presenting my own film *Weemoed & Wildernis* (2010). I initially thought, the Architecture department, they will have all kinds of technical questions about how many square meters, these kinds of things. So, I had prepared myself a little bit for such questions, but it was not at all like that. Instead, it was about wilderness, Dutch wilderness. There were people coming from Greece who told me they don't have a special word for wilderness because for them, everything is wilderness and

cultivated land is something special. So they have names for that, but not for the rest. That's the normal thing. That's of course just the opposite in the Netherlands. So, what is wilderness in the Netherlands? What kind of wilderness do we have here, if everything is constructed? For me, wilderness is nature taking over.

What do you think about the experiment of non-filmmakers using film as a way to bring together different disciplines?

That's great, that you can do this. Why not? I see that only in a positive way. I think most of my colleagues wouldn't mind. I'm doing the same thing: I also use archival material, which was not at all meant to be used the way I use it.

Anything else that you would think you want to add to our conversation?

Let me see. Awareness, we talked about it. Also in politics; that's what the film *Onder de oppervlakte* (*Below the Surface*; 2015) is about. Maybe someone's viewpoint is right, but to get your point across is difficult when trying to organize nature or to organize regulations on nature. So you have to find people to support your policy. But to get those people interested, you need that awareness. Democracy only works when people are well-informed and know what to decide and how to vote. Nowadays that's sort of a problem, I think.

Do you think we have too much information?

Yes, but also false information. How can we organize that? How can we gain trust again in the really important things? And how do we know what to trust? We need to be precise on what the sources are. I still think that science is not only an opinion. Science is real science, and that it is based on facts and on research, and not only on what we feel it is. Nowadays, this has changed, and it is very frightening. Activism, yes. Places are very important to me, yes. And memories.

References

Pamuk, O. (2008), *Museum of Innocence*? (New York: Vintage).

Sinke, D. (director, 2010). *Weemoed & Wildernis / Wistful Wilderness* [Film]. SNG Film.

Sinke, D. (director, 2015). *Onder de oppervlakte / Below the Surface* [Film]. SNG Film.

Sinke, D. (director, 2018). *Bewaren – of hoe te leven/Keeping and Saving or How to Live* [Film]. De Familie Film & TV.

Van Gasteren, L. (director, 1960). *Een nieuw dorp op nieuw land / A New Village on New Land* [Film]. Spectrum Film.

Van Gasteren, L. & Meerman, J. (directors, 2012). *Terug naar Nagele / Back to Nagele* [Film]. Spectrum Film.

Een nieuw dorp op nieuw land, Louis van Gasteren, 1960 + *Wieringermeer*, Alex Roosdorp, 1938

Rommy Albers

Een nieuw dorp op nieuw land (*A New Village on New Land*) is a documentary commissioned by the Ministry of Housing and Construction about the creation of the village of Nagele in the Northeast Polder, one of the reclamations in the IJsselmeer. Nagele is one of the smaller centres in the new polder. A number of architects and urban planners, like Aldo van Eyck, Cor van Eesteren, Gerrit Rietveld and Jaap Bakema, drew the plans for the new residential community, combining old ideas with new insights in modern architecture and urbanism: a central core with a church and shops, surrounded by a sheltering hedge of willow. Exemplary in this process is the move of a farm worker from his home near Kampen to a modern house on the new land. The film exudes the typical atmosphere of reconstruction after World War II; the optimism of the 1950s and early 1960s to rebuild the Netherlands in a concerted effort after the German occupation.

In 1938, agricultural filmmaker Alex Roosdorp had already shot footage in another reclamation area: the Wieringermeer. Roosdorp did not focus on infrastructure, but entirely on agriculture, in this case the grain harvest. The images he shows are un-Dutch: instead of small plots still being worked with horsepower and scythes, we see large fields of wheat from which the harvest is brought in with combines. This model farm is more reminiscent of the large granaries in Canada or the Soviet Union, but certainly does not fit the image of Dutch agriculture in the 1930s. Land extraction and reclamation brought great changes; new social engineering with modern village- and town planning and large-scale and mechanical tillage of the land. The new world presented itself.

In 2012, Louis van Gasteren made a sequel to his film: *Terug naar Nagele* (*Back to Nagele*). He talked to architects, planners, architectural historians and residents about present-day Nagele, and discussed whether the village still meets the demands of the present or whether social engineering turned out to be a utopia after all.

Dialogue with Lia Franken-de Vries

Lia Franken-de Vries and Carissa Champlin

Apparently, there is a lot of trust in the knowledge and capability of the Dutch in keeping their feet dry.

Dutch folklore tells of an age-old struggle to hold back the water. Many of these stories lay dormant, submerged under the tidal current of everyday life. In this remote, asynchronous interview with Lia Franken-de Vries, reflection surfaces early memories of disaster tourism and visiting the Delta works under construction. Later as alderwoman by the Municipality of Hollands Kroon, North Holland Province, Mrs. Franken-de Vries shares stories of a new adversary: the data centre. In the context of scarce freshwater resources and growing demand for energy production, these Dutch polder regions are feeding the world's appetite for data storage at the cost of the resource needs of local people, livestock and nature. Is this a new Dutch struggle in the making?

Growing up in the Delta

Carissa Champlin: Living with water was and is still a hallmark of life in the Netherlands. You were born and raised in Bergen op Zoom, a city located in the Brabant Province directly next to Zeeland. What are your earliest memories of living with and next to the water in your city? Was dealing with water a part of your daily life?

> Lia Franken-de Vries: Bergen op Zoom sits higher than the surrounding villages and I cannot recall that water played a significant role in my youth. We did however take regu-

lar weekend trips to the Schelde to collect cockles. We would camp in a caravan in Zeeland, but the campsite was protected by high dunes.

You were born a couple of years after the 1953 flood disaster. Was there a lot of talk about the disaster when you were a kid? What sort of after-effects did the disaster have on your city?

I was born 2 years after the disaster and when I was old enough to understand what people were talking about, already quite some time had passed. People did not speak much of it. This could be because we did not lose any family members or homes during the disaster. My uncle lived in Vlissingen and his family was evacuated. But he was able to return to his house later. When we would visit, we would look at how high the water stood. This marking was still visible on most homes. Other than having evacuees in the city, as far as I know, we did not experience any further effects.

The film *De lage landen* (*Hold Back the Sea*; Sluizer, 1960) details the large programmes to build the flood defences as quickly as possible. What did you see related to the construction of the Delta Works in your surroundings or on the news?

There were many reports on the progress of the Delta Works in the news programmes on tv. But don't forget that in the 1960s the news outlets were not as extensive as today with social media. I can remember that we would sometimes go by bicycle to look at the Works.

In what ways did the Delta Works change your everyday life and society in and around Bergen op Zoom and nationwide? Did older generations speak much about what it was like before the mega water projects?

The Delta Works did not change my daily life in any way. But they did have consequences for the fisheries and others. Eventually, there was only 1 anchovy fisherman in Bergen op Zoom because the water behind the Delta Works by and large became brackish. The older generation did not tell any significant stories about life before the Works.

Living in a protected 'polder' area

More than half of the Netherlands is vulnerable to flooding. You were an alderwoman for 12 years in Hollands Kroon. A significant portion of the municipality lies between 2 and 5 meters below sea level. At the same time, North Holland is optimally protected against flooding by its natural dunes along the coastline and its large waterworks such as the Afsluitdijk that spans 32 km, closing off the IJsselmeer from the Wadden Sea. Regarding water and flooding, what were the most important subjects for your constituents?

> Hollands Kroon is primarily an agricultural municipality. It was and still is important that water is drained in a timely manner during heavy rainfall events. In addition, retaining sufficient water in dry periods has also become increasingly important. It is also important for agriculture that 'salty seepage' is prevented as much as possible. This is bad for the irrigation of crops and the drinking water of livestock.

During your time as alderwoman, several large projects were implemented in Hollands Kroon including the new housing district De Sluis and Google and Microsoft data centres. What considerations did these businesses, and the local residents and farmers find important with regard to life in the polder area? What were the main topics of conversation in relation to safety and flooding in and around these large projects?

> The arrival of the data centres and the large wind farms was not met by everyone with enthusiasm. There were people who understood that the polder area was a suitable location, but others thought that they caused the landscape too much damage. We all use energy and data, but we also all struggle with the 'NIMBY' [Not in my backyard] effect. The most important considerations for the businesses were to have sufficient access to (wind)energy and water in addition to a calm social climate. Costs no doubt also played a role. It never ceased to amaze me that such large investments were being made in an area that lies 5 meters below sea level. Apparently, there is a lot of trust in the knowledge and capability of the Dutch in keeping their feet dry.
>
> Now that there are more and more data centres arriving, pushback from residents is growing. In projects still to be realized, self-sufficiency in the required energy and water supply will surely become a larger part of the considerations. To my knowledge, flooding or flood risk has not been a significant part of the discussion, perhaps due to the trust factor I mentioned earlier. Or that these discussions already took place at an earlier stage with Google and Microsoft before the projects were implemented.

What new developments have happened within the municipality since you stepped down as alderwoman? And to what extent do considerations over nature, climate and safety in the face of climate-related disasters play a role? Do you have the impression that priorities have changed since you were alderwoman or perhaps since you were growing up in the delta?

> Since I stopped serving as alderwoman in 2014, more data centres have been constructed. At the same time, the discussion has begun on whether this sort of decision can be left to an individual municipality or whether the province should play a larger role. Since the construction of the first data centres, more and more attention has been paid to their integration into the landscape and the demands placed on water and energy. Especially in consideration of drought: everyone is asked to be conservative in their water use, but does that also apply to cooling water, etc.? Or do companies have to make provisions for this themselves?
>
> Climate change and the continued subsidence of our soil will certainly become more important in the discussions. Not only the danger of sea level rise, but also rainwater, which is far less manageable, that enters our country via the rivers will require us to look differently at large projects or housing construction. We in the Netherlands have always been good at draining water as quickly as possible, but now ingenuity is also required to absorb heavy rainfall and to retain rainwater for dry periods.

References

Sluizer, G. (director, 1960). *De lage landen / Hold Back the Sea* [film]. Shell Film Unit.

The Edges of the Centre

Marilena Mela

This essay examines how contemporary critical environmental and decolonial frameworks and their emphasis on local knowledge can be relevant to European landscapes. This question arises in response to today's multiple crises, which make it evident that the focus on progress and economic development, supported by geoengineering or terraforming, accelerates environmental catastrophes and reproduces social injustice. Many scholars trace the historical roots of the climate crisis to Western modernity, with colonialism, imperialism and capitalist expansion bringing prosperity to European landscapes, leading to extraction, exploitation and destruction in other parts of the world. Now European landscapes also face the consequences of planetary ecological degradation. In light of this crisis, critical and decolonial frameworks point toward alternative ways of relating to the environment, including pre-modern, non-Western and local perspectives. How can these perspectives stay relevant for European landscapes without overshadowing the historical and systemic unevenness entailed in their shaping? I examine this question by focusing on a well-studied edge of Europe – the Dutch coast, specifically the Wadden Sea area – exploring the potential of theory to build reparations, resistance, and solidarity, and discussing its implications for spatial practices.

Beyond Geoengineering

This essay is an effort to bring different worlds together. Reporting the Delta examines the coast of the Netherlands and the legacies of planning and engineering works. It also explores landscapes and other ways of understanding their past, present and future – a need made urgent by the accelerating environmental crisis. These 'other ways' are my focus here. To explore them, I draw on theories and approaches by thinkers who have examined how the painful consequences of Western colonialism and globalized capitalism in non-European places – often in (post)colonial contexts – give rise to paradigms of resistance, environmental caretaking and co-creation, building on systems of local knowledge. As the planetary crisis makes collective action necessary, how can the study of the making of European landscapes be situated within this discussion?

In the Netherlands, the transformation of territories at the interface of land and water is a central theme in virtually all historical accounts. Petra van Dam describes Dutch culture as an 'amphibious' one, linking memory and culture with environmental history (van Dam, 2016). The numerous small stories of environmental adaptation by communities, which transformed landscapes in order to dwell, grow food, exchange goods and socialize, are central in any regional imagination in the Netherlands (Ritson, 2018). Over time, these small-scale adaptations were supplemented or replaced by larger projects, including land reclamation, industrialization and resource extraction. This shift occurred over an extended period, beginning with advancements in engineering and technology in the late Middle Ages, accelerating with the integration of many regions into the market economy in early modern times, and further intensifying with the industrialization of the late modern era (Lotze et al., 2005). It is in this period that landscape transformation took on the characteristics of rigorous geoengineering – terraforming on a scale and intentionality that led to the country being described as 'a self-made land' (van der Cammen et al., 2012). After 1960, the persistent claim to protect and preserve nature has made nature preservation an equally potent agent in the making of this land.

All of the above is well-known, and the reader might wonder why I offer this elementary interpretation of the making of Dutch landscapes. The reason is that I aim to tackle a conceptual tension. In the Dutch context, accounts of geoengineering connect to narratives of progress and development driven by the ingenuity of Dutch spatial planners. Indeed, the contributions of monumental spatial planning and engineering projects to safety and quality of life are unquestionable. However, contemporary thinkers in critical sectors of the humanities and social sciences see 'geoengineering' or 'terraforming' as typical practices of modernist, colonial and imperialist expansion historically led by 'the West' – processes associated with catastrophe in other parts of the world, and eventually for the entire planet. Indeed, Amitav Ghosh traces the origins of the climate crisis to colonial expansion, which, through its political and financial dominance, established certain ways of understanding and shaping the environment – including geoengineering – and erased others (Ghosh, 2022). The painful histories of the extermination and exploitation of people, landscapes and lifeworlds in colonized lands are dark legacies of this history. The change of climate and its widespread consequences are also legacies of the same system, affecting Earth and all her inhabitants. Existing responses to climate change remain inscribed within similar frameworks, defined – as Donna Haraway puts it – by 'technotheocratic geoengineering fixes and wallowing in despair' (Haraway, 2016). Meaningful responses should instead echo the call for ontological politics, as expressed in the work of Isabel Stengers. Stengers argues that the global West and its hegemony of 'progress', 'rationality' and 'objectivity' became 'machines' that destroy 'both ontologies and politics' (Stengers, 2018, 86). Such crimes have destroyed many worlds, and politics ultimately has the collective responsibility *to resist being destroyed.* It thus becomes imperative to look for situated, living or latent alternatives to destructive systems.

In this context, thinkers direct attention to landscape practices that might point towards other pasts and futures for humans and nonhumans that inhabit them. Local knowledges of the land, the water, the mountain, the river, the volcano, exist as drivers of inspiration, resistance and resilience (Bocco Guarneri & Habert, 2024; Harvey, 2015). These can be posed against imaginations about the planet as a set of resources that can be tamed, tailored and engineered. Paying attention to these 'alternatives', both living and historical, is crucial for two reasons: first, they uncover power imbalances that originate from centralized systems of seeing, planning and governing, which exploit landscapes and disregard the vitality of communities. And second, they show the way toward systems with the potential to provide answers in this epoch of despair.

Let me connect these insights to the Dutch coast. In his work about climate change, Bruno Latour argues that 'in the landscape, we find a place to land', meaning that local landscapes make abstract global challenges pressing and urgent and can act as anchors of empathy (Latour, 2018). Conversely, the histories of Dutch landscapes must be reconnected to a critical history of the planet. The reality of the climate emergency means that any change of approach needs to assume planetary and collective dimensions, with simultaneous resistance to extractive agendas and new 'designs' in different landscapes (Escobar, 2018). In this context, the emphasis on alternatives to growth and expansion is relevant for all landscapes, including those situated in the global North and West, and particularly in peripheral, rural or economically marginalized places in these geographies. Of course, this is a dangerous conceptual endeavour: whatever the extent of their peripherality in their national contexts, the well-being of European populations and landscapes remains historically tied to the ending of worlds elsewhere. Building resistance, knowledge exchange and solidarity in face of planetary threats must coexist with the recognition of this unjust imbalance and resist its reproduction. This is why critical environmental discourses must remain paired with decolonial frameworks.

The second reason is that research on local knowledge in the context of Europe rarely integrates insights from decolonial thought. In the face of climate change, scholars and practitioners study systems of traditional climate adaptation for their potential to address contemporary problems. In (post)colonial and indigenous contexts, research has often reproduced the marginalization of local communities, treating their knowledge as a resource to be extracted. Critical scholars and activists argue that these knowledges cannot be detached from the social and political realities in which they were created, nor from broader political claims (Smith, 2022). In Europe, such research remains depoliticized: the 'vernacular' and 'local' are either romanticized as something irrelevant to today's struggles, or reduced to merely technical perspectives (Bocco Guarneri & Habert, 2024). This gap highlights the necessity of learning from broader emancipatory discourses. Local knowledges are crucial not only because they offer technical or engineering solutions, but also because they reveal different ways in which people have socialized, built, produced and governed land, as well as the contemporary continuities of these models. In other words, this perspective must prioritize both local diversification and global political urgency.

Can we then view places in the Netherlands – a country which was historically involved in colonial practices and expansion – as having the potential to strengthen bonds between people and landscapes and offer alternatives to centralized geoengineering? What historical processes and current local needs make this endeavour necessary? And how can thinkers and practitioners involved with Dutch landscapes learn from critical environmental and decolonial discourses?

Localizing the Debate

In the past few years, my research has brought me back time and again to the Dutch part of the Wadden Sea area. Drawing on my familiarity with the region, I will outline a few characteristics of its landscapes to illustrate how the above arguments can be localized.

The Wadden Sea is a mudflat area that submerges during high tide. The sea and its adjacent lands fall within the provinces of North Holland, Friesland and Groningen, and extend into Germany and Denmark. It is a changing territory where sandbanks, inlets, shoals and islands constantly emerge or disappear due to the forces of tides and currents. The history of human habitation is characterized by unique examples of adaptation to the environment, starting 7,500 years ago, when the first hunter-gatherers reached the area. Landscape transformations first intensified 2,500 years ago, including the use of ditches, small dikes and dwelling mounts (*terpen*) that remained dry in times of flooding (Bazelmans et al., 2012). In the past one-and-a-half century, monumental defence works mitigated the centuries-old sense of risk that is interwoven with the area's sense of place (Bankoff, 2013).

In the same period, and especially the second half of the nineteenth century, investors and developers saw land consolidation, industrial fishing, gas drilling and other infrastructures as ways to optimize the use of the area's resources. A series of geoengineering projects took place that were typical for the Dutch spatial planning culture of the time. Examples are the plans for the reclamation of the entire Wadden Sea as agricultural land in 1871, through the construction of a seven-kilometre dam (Figures 1 and 2); this plan was abandoned after a storm (Schroor, 2018). In 1920, further plans for the draining of the sea as part of the construction works of the Afsluitdijk and the reclamation of the Zuiderzee by Cornelis Lely also remained unrealized. In 1974, J. Thijsse proposed the partial reclamation of the sea, which would be converted into a lake (van der Cammen et al., 2012). This last endeavour roughly coincided with the discovery of gas in the Wadden Sea area and in Groningen. In Ameland, one of the Wadden Sea islands, a heated debate took place between 1964, when gas was first found, and 1981, when the gas drilling infrastructures were put into use (Figure 3) (Hakkenes, 2020). Multiple studies confirm that biodiversity losses and destruction of landscapes occurred in this era as a result of these infrastructural projects (Lotze et al., 2005).

These large-scale projects of land consolidation and gas drilling were met with intense reactions from environmental activists. After 1965, their actions managed to bring about a change of paradigm in the management of the Wadden Sea area (Schroor, 2018). It was in this period that the sea came to be seen as a valued natural landscape, a process culminating in a trilateral World Heritage nomination of the area, extending from the Netherlands to Germany and Denmark. This dominance of the importance of nature preservation came not without friction: local communities protested the loss of autonomy over the territory and its separation into strict zones of protection and development. In proclaiming the sea protected nature, its boundary with the adjusting land became harder and more definitive (Egberts, 2019). Cultural geographers have argued that the longue durée of dwelling in the Wadden Sea area was instead characterized by a more porous interface between land and water that took into account temporal landscape dynamics, unpredictability and risk. Borrowing from the archaeologist Christer Westerdahl, and in contrast with the Wadden Sea's popular representation as a primarily 'natural site', Hans Renes describes the region as a maritime cultural landscape and urges to take its cultural history into account in research and spatial governance (Renes, 2018).

The past of Wadden Sea, where humans have consciously transformed the environment for millennia, makes it a 'predicament of the Anthropocene' (Ritson, 2018). At the same time, its spatial history is characterized by planning and management approaches based on modern principles of objectivity and rationality, expressed in zoning and strict boundaries (Fainstein & DeFilippis, 2016). Nature preservation has undoubtedly protected the ecosystem of the Wadden Sea. Still, it has been less successful in engaging locals in its management, thus reinforcing the separation between humans and nature (Descola, 2017). My interest lies in the ways centralized approaches coexist with other diverse systems of thought and practice, originating in the long past of landscape making.

First, we see people understanding the dynamic nature of their environments and finding ingenious ways to live with them. Take, for example, the creation of dwelling mounts; the multiple types of dikes, first around villages or farms, later on enclosing larger areas; the creation of polders and drainage systems; the development of solutions for transport via water; the assembling of construction materials, often through recycling and reuse; the hybridity and periodicity of using saltmarshes, peat bogs and reed beds; and the engineering of ditches, pumps, windmills and polders (Schroor, 2018). Wind, rain and snow in all their unpredictability; the tides, the soil composition and the movement of sandbanks play roles in these adaptations. To understand them presupposes skill, craft, perception; in short, knowledge. But this is place-specific knowledge, one that is only transferable through experience and direct engagement. Déri and Sundaresan see local knowledge as connected to spatially and temporally defined communities that transfer it in intra-generational ways (Déri & Sundaresan, 2015; Geertz, 1983). As external observers, we may not be able to capture this knowledge. Still, we can sense its presence through the many spatial traces of the past and the continuation of traditional practices (Figures 4 and 5).

Second, learning from decolonial approaches means looking at the politics of land, its distribution, governance and management, and how these politics might differ from modern conceptions of public administration and planning. One example is the existence of systems of common land ownership and use. In Ameland, which remained an independent lordship between the fifteenth and the nineteenth century, an agreement allowed islanders to regulate land use in a system of commons. Land was divided into parts, each with different ownership status and land use (Brouwer, 1936). Each farm made use of three types of land: sandy soils in the dunes were used as arable land; the peaty soils of the embankment were used for pasture under a system of periodic subdivision; the land outside the dikes remained undivided and was used as common pasture. The Mark, the local assembly, was responsible for regulating rights to its use and for overseeing roads, ditches and dikes. Brouwer sees this as a necessary system of social organization in response to drought, flooding, inundation, drifting sands and the shrinking of land, challenges that 'made it impossible for individuals to protect against such threats alone' (Brouwer, 1936, 30). The Mark of Ameland was dissolved at the end of the nineteenth century, replaced by more central models of administration. Common lands were enclosed and subsistence activities gradually gave way to tourism and recreation. Jan van Zanden, discussing the dissolution of the marks in eastern Netherlands, observes how their existence might have favoured landless farmers or micro-ownership, and made it possible to avoid the overexploitation of land under specific conditions (van Zanden, 1999). The water boards, which in some cases replaced the Marks, are another form of socio-spatial organization based on collaboration and caretaking. Their authority has been increasingly centralized and institutionalized, but historically, they were closely related to local conditions and local governance. The political underpinnings of these governance structures are examined by Kenneth Olwig, who has shown how the Dutch word *landschap* entails this political and territorial understanding of landscape (Olwig, 1996).

Third, seemingly intangible aspects – traditions, symbols, legends, rituals – are also linked to environmental conditions and imaginaries. In the case of the Wadden Sea area, which largely belongs to the region of Friesland, a strong regional identity is based on the history of the Frisian kingdom and the use of a distinct language. Stories of origins are interwoven with a consciousness of the local landscape that engages with environmental events and memories, including the deification of the Sea, tales of drowned villages, commemorations of floods and celebrations of local ingenuity. Legends such as Rixt van Oerd from Ameland, an old woman who lived in an unembanked area collecting driftwood and herbs and communed with the devil, or Ame Gijs, a spirit who lived by the water and threatened children, are personifications of the wildness of an ever-changing and often dangerous landscape. In her work about literary imagination in the North Sea, Katie Ritson argues that stories show how space is invested with meaning, which makes it possible to reflect on humanity's place in the world in the face of human engineering and ecological vulnerability (Ritson, 2018).

The three previous paragraphs sketch an image of the making of the Wadden Sea landscapes as space, politics and narratives. Local, vernacular, pre-modern, indigenous knowledges are surely multiple, interconnected and linked to changing cosmologies. However, in contemporary formal knowledge systems, the three approaches – spatial adaptation, systems of governance and storytelling – have been compartmentalized. They are viewed separately as either technical expertise, historical accounts or folklore and tradition, each falling into different domains of science or policy. A look at indigenous systems might be helpful in shifting this perspective: spatial adaptation, governance and cosmologies in their pre- or anti-modern state are not separated, but are interconnected aspects of holistic understandings of the world (Harvey, 2015). Even though science and policy tend to compartmentalize them, we can hypothesize that the sense of place and local knowledge maintains this holistic nature as long as people continue interweaving their livelihoods with landscapes.

Rescaling Is Not Upscaling

All this shows how this telling of the Wadden Sea's history connects to a broader critique of modernist ecological imaginations. Even within a 'developed' country such as the Netherlands, economic peripheries and cultural margins persist. The destruction of ecosystems and the extraction of resources are linked to narratives of exploitation and land dispossession. Even celebrated instances of spatial planning and nature conservation originating in those same systems of thought have often excluded humans and nonhumans from their home landscapes through the imposition of strict boundaries and realms of authority. I see such places as 'the edges of the centre', edges that can both benefit from and enrich critical environmental approaches.

I discussed the Wadden Sea landscapes as made, given meaning to, governed and reenacted collectively and intertemporally in the longue durée of human dwelling. As heritage scholars David Harvey and Jim Perry propose, a creative engagement with the spatial history of the area shows how communities have navigated adaptation, loss and creativity in response to changing environments and climate conditions (Harvey & Perry, 2015). Local legends and celebrations, corresponding to environmental imaginaries, connect to what Amitav Ghosh calls 'vitality', showing landscapes and nonhumans coming to life through communal songs, performances and stories (Ghosh, 2022). The knowledge of the land and sea, expressed in continued hybrid practices of farming, pasture and building, demonstrates its value as 'pedagogy', as described by Michi Saagiig Nishnaabeg scholar Leanne Betasamosake Simpson, who sees grounding knowledge in place as a pathway to resistance and resurgence (Simpson, 2014). Common ownership, use and governance of land – which is frequent in indigenous societies and part of the living history of Dutch coastal landscapes, even if it has been dismantled – link to political claims about local communities' subsistence, agency and autonomy (Loeffler, 2021). These insights make it possible to see local knowledge of the Dutch coast as tied to an alternative human-landscape coexistence that has been ruptured but not completely broken.

The takeaways of this retelling of history for researchers and practitioners in Dutch landscapes can be multiple and multiplying. Below, I outline four lines of implications.

First, efforts have already begun toward hybrid, land-based environmental solutions that transcend hard geoengineering and embrace the knowledge of local landscapes. Even land defence projects often go beyond rigid zoning and overly technocratic solutions, prioritizing hybrid designs and dynamic land management. For instance, the Living Dikes project experiments with nature-based and climate-responsive alternatives to conventional dikes (Borsje et al., 2025). Other examples in the Dutch context, spanning various scales, include nature preservation and biodiversity, food production and landscape adaptation. What is urgently needed is not only to multiply such efforts but to connect them within broader frameworks that transcend disciplinary boundaries.

Second, prioritizing local knowledge, both in research and governance, must go hand in hand with prioritizing local people. In the Netherlands, governance has indeed shifted toward participation, democratization, community empowerment and dethroning the expert, particularly in 'softer' fields like heritage (Janssen et al., 2017). However, more is needed. We need to actively learn from scholars outside Europe, where systems of oppression and disempowerment tied to research apparatuses have long been critically examined (Smith, 2022). Researchers, practitioners and policymakers must take local communities seriously as knowledgeable and creative assemblies that govern their inherited landscapes under principles of accountability, collaboration and sustainability.

Third, efforts like Reporting the Delta itself showcase the power of arts, through cinema, literature, painting and other mediums, to illuminate and critique cultural imaginaries that resist hegemonic paradigms. As this book demonstrates, films and documentaries have long engaged with the lifeworlds intertwined with geoengineering projects and shown the need to politicize debates surrounding them. The unveiling of modernist troubles comes through the approach of landscape as a holistic milieu that constantly evades rigid boundaries and categorizations. Such means also have the potential to make people feel, and eventually act, as being led by empathy and creativity.

Fourth, the history and future of Dutch landscapes must be situated within the broader history and future of the planet. Critical environmental discourses cannot be divorced from decolonial claims – both for reasons of social justice and because of the sharedness of planetary challenges. We must first acknowledge that constitutional structures and epistemologies of Western modernity can be traced to knowledge situated not far from the Wadden Sea. Immanuel Wallerstein has illustrated how the Dutch expertise in taming their landscapes enabled their ascent to global dominance in the seventeenth century, leaving painful legacies of colonialism and the slave trade (Wallerstein, 2011). Recognizing this history demands solidarity with the resistance of underrepresented or marginalized communities in (post)colonial geographies. This needs to become a firm basis for any new paradigm of planetary stewardship.

In this essay, I explored possibilities for linking critical environmental and decolonial discourses with the future of European landscapes, using the Dutch Wadden Sea region as an example. I recognize that I have employed terms such as 'the West', 'modernity', 'decolonial discourses' and 'local knowledge' somewhat monolithically, which risks oversimplifying their dynamic coevolution across scales and geographies. Nevertheless, I take this risk because I believe it is essential to connect previously disconnected domains to contribute to new alliances and forms of resistance. Once again, I argue that, in the context of the climate crisis, there are valuable insights in the relationships between people and their environments that offer alternatives to modernist models of expansion and the associated research, governance and planning frameworks. These answers persist in spatial traces, stories, governance models and latent practices – including those in landscapes of so-called 'developed' countries, where struggles might stem from the same modernist infrastructures. Rescaling our attention to local landscapes, their struggles, agency and embedded knowledge does not mean scaling up to new universal systems. The political potential of this rescaling, however, cannot be fully activated without an active recognition and unmaking of historically rooted unequal geographical relationships and the building of networks of empathy and solidarity. Instead, we must not only make space for different worlds to thrive in their diversity, but actively work to interconnect them in pursuit of larger, urgent claims.

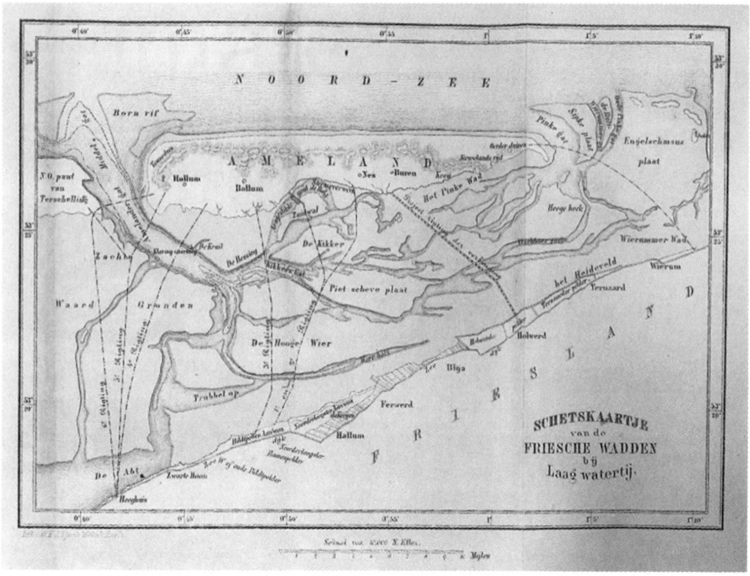

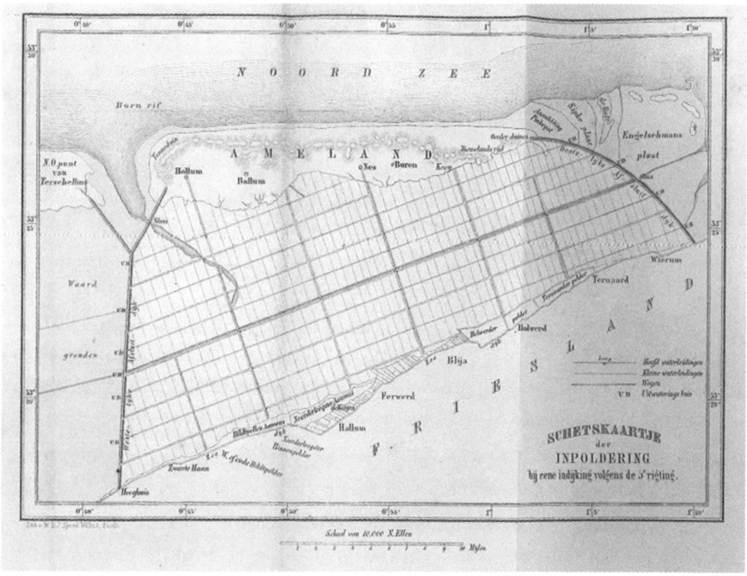

Figures 1 and 2: Plan of 1877 showing the proposed dam between Ameland and Holwerd and the projected results of the land reclamation. Amelander Historie, 2024.

Figure 3: In 1897, state tax collector Jan Gabes Scheltema (centre) was photographed on the partially constructed, now-defunct dam to Ameland, which was intended to connect the island to the mainland as part of a land reclamation attempt. Asing Walthaus, *Leeuwarder Courant*, 21 August 2020.

Figure 4: The NAM gas drilling platform, visible from the beach on the eastern part of Ameland, 2023. Photo: Marilena Mela.

Figure 5: Where the embanked and unembanked parts of Ameland meet, 2023. Photo: Marilena Mela.

Figure 6: Exhibition in Landbouw-Juttersmuseum Swartwoude, dedicated to traditional activities, including transport, farming, and animal husbandry. Arable cultivation has largely ceased, but a significant part of the island continues to be used for pasture. Ameland, 2023. Photo: Marilena Mela.

References

Amelander Historie (2024). *Over Ameland en 'daar ligt de dam', van plannen die niet doorgingen*. Available: https://www.amelanderhistorie.nl/news/over-ameland-en-daar-ligt-de-dam-van-plannen-die-niet-door-gingen/ (accessed March 2024).

Bankoff, G. (2013). 'The 'English Lowlands' and the North Sea Basin System: A History of Shared Risk', *Environment and History, 19*(1), 3-37.

Bazelmans, J., Meier, D., Nieuwhof, A., Spek, T., & Vos, P. (2012). 'Understanding the Cultural Historical Value of the Wadden Sea Region. The Co-Evolution of Environment and Society in the Wadden Sea Area in the Holocene up until Early Modern Times (11,700 BC-1800 AD): An Outline', *Ocean & Coastal Management, 68*, 114-126.

Blaser, M., & De La Cadena, M. (2018). 'Introduction: Pluriverse Proposals for a World of Many Worlds', in M. De La Cadena & M. Blaser (eds,), *A World of Many Worlds*, 1-22 (Durham, NC: Duke University Press).

Bocco Guarneri, A., & Habert, G. (2024). 'New Vernacular Construction: Environmental Awareness and Territorial Inclusivity', *IOP Conference Series: Earth and Environmental Science,* 1363. doi: 10.1088/1755-1315/1363/1/012114

Brouwer, D. A. (1936). *Ameland. Een sociaal-geografische studie van een Waddeneiland* (Amsterdam: Paris).

Déri, A., & Sundaresan, J. (2015). 'Diverse Epistemic Traditions in Transformative Climate Change Research and Adaptation: Heritage and Legacy', in D. C. Harvey & J. Perry (eds.). *The Future of Heritage as Climates Change: Loss, Adaptation and Creativity,* 78-94 (London: Routledge).

Descola, Ph. (2017). *Rethinking the Nature / Culture Divide.* Public Lecture at Chulalongkorn University. Available: https://www.youtube.com/watch?v=rVgUc2Rflfw

Egberts, L. (2019). 'Moving beyond the Hard Boundary: Overcoming the Nature-Culture Divide in the Dutch Wadden Sea Area', *Journal of Cultural Heritage Management and Sustainable Development, 9*(1), 62-73. doi: 10.1108/JCHMSD-12-2016-0067.

Escobar, A. (2012). *Encountering Development: The Making and Unmaking of the Third World* (Princeton, NJ: Princeton University Press).

Escobar, A. (2018). *Designs for the Pluriverse: Radical Interdependence, Autonomy, and the Making of Worlds* (Durham, NC: Duke University Press).

Fainstein, S. S., & DeFilippis, J. (2016). *Readings in Planning Theory* (Hoboken, NJ: Wiley-Blackwell).

Geertz, C. (1983). *Local Knowledge: Further Essays in Interpretive Anthropology* (New York: Basic Books).

Ghosh, A. (2022). *The Nutmeg's Curse: Parables for a Planet in Crisis* (London: John Murray).

Hakkenes, E. (2020). *Gas: het verhaal van een Nederlandse bodemschat* (Amsterdam: Thomas Rap).

Haraway, D. J. (2016). *Staying with the Trouble: Making Kin in the Chthulucene* (Durham, NC: Duke University Press).

Harvey, D. (2015). 'Landscape and Heritage: Trajectories and Consequences', *Landscape Research, 40*(8), 911-924. doi: 10.1080/01426397.2014.967668

Harvey, D. C., & Perry, J. (2015). 'Introduction: Heritage and Climate Change: The Future Is Not the Past', in D. C. Harvey & J. Perry (eds.). *The Future of Heritage as Climates Change: Loss, Adaptation and Creativity,* 3-21 (London: Routledge).

Latour, B. (2018). *Down to Earth: Politics in the New Climatic Regime* (Cambridge: Polity Press).

Loeffler, J. (2021). 'Restoring Indigenous Mindfulness within the Commons of Human Consciousness', in J. Hausdoerffer, B. P. Hecht, M. K. Nelson, & K. K. Cummings (eds.), *What Kind of Ancestor Do You Want to Be?*, 130-137 (Chicago: University of Chicago Press).

Lotze, H. K., Reise, K., Worm, B., Van Beusekom, J., Busch, M., Ehlers, A., Heinrich, D., Hoffmann, R. C., Holm, P., Jensen, C., Knottnerus, O., Langhanki, N., Prummel, W., Vollmer, M., & Wolff, W. J. (2005) 'Human Transformations of the Wadden Sea Ecosystem through Time: A Synthesis.' *Helgoland Marine Research, 59*(1), 84-95. doi: 10.1007/s10152-004-0209-z.

Moore, J. W. (2016). 'Introduction: Anthropocene or Capitalocene? Nature, History, and the Crisis of Capitalism', in J. W. Moore (ed.), *Anthropocene or Capitalocene? Nature, History, and the Crisis of Capitalism*, 1-13 (Binghampton, NY: PM Press).

Olwig, K. R. (1996). 'Recovering the Substantive Nature of Landscape', *Annals of the Association of American Geographers, 86*(4), 630-653.

Renes, H. (2018). 'The Wadden Sea Region as a Cultural Landscape, in M. Schroor & H. Renes (eds.), *Waddenland Outstanding: History, Landscape and Cultural Heritage of the Wadden Sea Region*, 45-64 (Amsterdam: Amsterdam University Press).

Ritson, K. (2018). *The Shifting Sands of the North Sea Lowlands: Literary and Historical Imaginaries* (Abingdon: Taylor & Francis).

Schroor, M. (2018). *Landschapsbiografie van het Waddengebied: Historisch-landschappelijke karakteristieken en hun ontstaan* (Amersfoort: Rijksdienst voor het Cultureel Erfgoed).

Simpson, L. B. (2014). 'Land as Pedagogy: Nishnaabeg Intelligence and Rebellious Transformation', *Decolonization: Indigeneity, Education & Society, 3*(3), 1-25.

Smith, L. T. (2022). *Decolonizing Methodologies: Research and Indigenous Peoples* (3rd ed.). (London: Bloomsbury Academic).

Stengers, I. (2018). 'The Challenge of Ontological Politics', in M. De La Cadena & M. Blaser (eds.), *A World of Many Worlds*, 83-111 (Durham, NC: Duke University Press).

van Dam, P. J. E. M. (2016). 'An Amphibious Culture: Coping with Floods in the Netherlands', in P. A. Coates, D. Moon, & P. Warde (eds.), *Local Places, Global Processes: Histories of Environmental Change in Britain and Beyond*, 78-93 (Havertown, PA: Windgather Press).

van der Aa, B. J. M., Groote, P. D., & Huigen, P. P. P. (2004). 'World Heritage as NIMBY? The Case of the Dutch Part of the Wadden Sea', *Current Issues in Tourism, 7*(4-5), 291-302. doi: 10.1080/13683500408667986

van der Cammen, H., de Klerk, L. A., Dekker, G., & Witsen, P. P. (2012). *The Selfmade Land: Culture and Evolution of Urban and Regional Planning in the Netherlands* (Amsterdam: Spectrum).

van Zanden, J. L. (1999). 'Chaloner Memorial Lecture: The Paradox of the Marks. The Exploitation of Commons in the Eastern Netherlands, 1250-1850', *The Agricultural History Review, 47*(2), 125-144.

Wallerstein, I. (2011). *The Modern World-System II: Mercantilism and the Consolidation of the European World-Economy, 1600-1750* (Berkeley, CA: University of California Press).

Art-Science Alliances and the Crisis of Authority

Jan-Philipp Possmann

Art-science collaborations have become increasingly popular in recent years, yet their promise often contrasts with the misunderstandings and misguided desires that can arise in practice. Drawing on a 25-year career in the cultural sector and extensive involvement in such projects, including both successes and failures, this essay reflects on the challenges of interdisciplinary collaboration. While artists and scientists share many similarities, their interactions are often hindered by idealized perceptions and a lack of understanding of each other's professional realities. Using Paul Feyerabend's concept of 'method pluralism' as a lens, the essay examines the dynamics of these collaborations in the context of a broader crisis of authority, exacerbated by the rise of populism in Europe and the USA. Through historical examples and personal experiences, it offers insights into navigating the complexities of art-science partnerships and fostering more meaningful exchanges.

Introduction

Working in European live art institutions in the 2010s was very exciting and inspirational, but also very challenging. In the institutionalized arts, the decade was shaped by institutional critiques, including excessive hierarchies, lack of transparency, little diversity among the staff, toxic work relations and inappropriate behaviours, unfair pay and, above all, worryingly homogeneous audiences. This all culminated coincidentally in the temporary closure of theatres and museums across Europe during the Covid pandemic in 2020, which felt like the coup de grace for a whole scene and an entire generation of artists.

I want to use some historical moments between 2010 and 2020 to illustrate how the cultural sector has tried to deal with these issues and what role academia and particularly art-science collaborations have played and continue to play in the continuous adaptation of institutional art and academic science to a changing socio-political environment. Two quotes from public figures of the decade seem to me to summarize well the similar yet different challenges the cultural and academic sectors are facing currently. The first is Dutch politician Geert Wilders calling art a 'leftist hobby' in 2010 (Fogteloo, 2010); the second one is Greta Thunberg calling on the UK Parliament to 'unite behind the science' in 2019 (Thunberg, 2019).

As a cultural professional with 25 years of work experience in various institutions, mostly in Germany, I will do so by referencing my own work as well as focus on the practical matters of working professionally in the cultural sector. This also reflects my conviction that many of the misunderstandings and misconceptions between artists, cultural workers and academic scientists that trouble art-science collaborations are due to a lack of knowledge of the practical work conditions, rather than philosophical or political differences. In fact, as I will argue, artists and scientists have a lot in common sociologically, but also in their self-images, doubts and aspirations. Yet, I suspect that both groups tend to see the ideal rather than the reality in the other. As someone actively working in and struggling with the institutional constraints of the cultural sector, I want to contribute to the debate by sharing my own experiences and mistakes in cultural production, addressing scientists in academia who are interested in art-science collaborations.

I use the term 'collaboration' instead of the more common 'cooperation', following Max Glauner's critical analysis of power relations in collaborative work relations: 'Collaboration is a mode of working together as equals: everyone has an equal influence on the process and the result. The starting point and framework conditions can be negotiated and redefined by all participants at any time' (Glauner, 2024). It should, however, be understood that this formulates an ideal rather than a factual description of the reality of many art-science encounters.

Scientists on stage in cultural institutions were not new around 2010, but this was definitely a growing trend. I had been actively producing and promoting art-science collaborations since 2009, when I produced an edition of Hannah Hurtzig's 'Black-market for Useful Knowledge and Non-Knowledge', an event in which visitors could book a 30-minute session with an expert of their choice. From 2010, I created and curated together with Adrienne Goehler a successful live talk show series at Nationaltheater Mannheim, devoted to encounters between culture, science and politics. When the series concluded after 5 years, the theatre transitioned to a new format, replacing our informal, niche salon with a more prestigious and authoritative lecture series.

By 2015, lecture formats on theatre stages as well as online had become hugely popular. Videos of TED Talks became available online for free in 2006 and by 2011, TED Talks' combined viewing figures surpassed 500 million (Kessler, 2011). The success of TED Talks online and of TEDx live events illustrates well the ambivalence of the popularization of academic formats at the time. On the surface, TED Talks seem to pay tribute to the centuries-old lecture hall of academia. Consequently, from an aesthetic point of view, they are rather sparse and comparatively dull. What made them attractive for a non-academic crowd, I would argue, is that they were in fact not paying tribute to the academic system of validation but rather to innovation, originality and most of all the charisma or performative skill of staging. The TED Organization describes itself as 'devoted to curiosity, reason, wonder and the pursuit of knowledge – without an agenda. We welcome all who seek a deeper understanding of the world and connection with others, and we invite everyone to engage with ideas and activate them in your community' (TED Organization, 2024). Clearly the non-elite, non-expert, egalitarian appeal is TED's selling point. By popularizing the academic formal canon, these shows in fact fundamentally question its value system. Academic merits, long-earned expertise and scientific rigour are replaced by inspiration, emotional authenticity and daring – qualities we usually seek in and attribute to artists and art works.

As I will show in this piece, TED Talks can be regarded as a good example of how academia and art have borrowed formats and discourses from each other in order to find ways to meet the demise of social authority which both have been faced with over the last two decades.

What Is Art Anyway?

So, if TED Talks are not really science, are they art? Maybe the most fundamental question scientific institutions should ask themselves before engaging in art-science collaborations is this: what is art, or even better: what do we want art to be? The first place to go might not be an artist union or a museum but an art historian in a university.

Unfortunately, the question 'what is art?' has as many answers as there are philosophy professors in Europe. For the purpose of this article, let's stick to the basics: I think it's vitally important to differentiate between art as a profession and the many ways in which people refer to aesthetic experiences and aesthetic qualities in their everyday lives. Much to my surprise, this is still one of the most common misunderstandings I have encountered outside of my profession. While everybody has the urge to express themselves and everybody has a sense of beauty and of the cultural – that is aesthetic – agreements within her or his social environment, we are not all artists. Obviously most scientists know this, but there is still surprisingly little insight and even interest in the methods and structures by which art as a profession is performed.

I assume that this is the fault of the artists themselves, who at least since the Romantic period in the early nineteenth century have invested a great deal of their craft into obscuring these structures and conditions. Why? It must always have been a very successful marketing strategy. Since the development of what Theodor Adorno called the 'administered world' (Adorno et al., 1989), there is a deep yearning in modern societies for life outside of these restrictions and regulations – let's call it the capitalist utopia. Artists have successfully claimed a kind of gatekeeping position for this utopia, rivalled only by drug dealers and maybe children. Part of their strategy has been to obscure the professional, more structured side of what artists do.

My definition of professional art might seem rather dull, compared with the popular discourse, but it has served me very well in 25 years of work in the field. Professional art in the European tradition is an experimental study of forms of communication. Put differently: artists investigate, disassemble and reassemble forms, sometimes creating new ones that eventually become used by others. And since we live in capitalist societies, where contemplation is valued very little and consumption very much, form is not something many people spend time wondering about. This definition of art also has the advantage of assigning a clear social functionality to it: applied science of communication. While you may argue that any good researcher in any academic field should have an awareness of language, imagery and physical performance, there is a big difference between using stylistic devices to improve your argument and to fundamentally question the style. I believe that a rigorous, profound and critical questioning and reformulation of form is essential for democratic society. The problem is of course that any research is already informed in that sense. That's why the job cannot be assigned to, let's say, linguists or communication researchers in a university. When playing around with form, profoundly, you are bound to confuse people. And we need professional institutions and spaces designated to this confusion: the museum, the theatre, the concert hall.

If we see artists as applied scientists of communication, the gap between academia and art of course narrows significantly. Like science, art is a specific, highly specialized profession, developed over time. As such it has rules, methods, limitations and working hours, as I will discuss below. Showing an interest in them and understanding them as much as possible is crucial for successful collaboration, no matter how much display of unprofessionalism or extravaganza the artist puts on. Like all professionals, artists know how to play their cards well and the artist's trump card is unconventionality.

Freedom within Tradition

The historian of science Paul Feyerabend used the term 'tradition' to describe the various ways of knowledge production. I find the term useful as it highlights the historical aspects of the way we try to make sense of the world and find solutions for problems. A tradition is not something you just decide to do or a way to do it, but is something you find, investigate and put to use. By adopting the tradition of scientific rationality or the tradition of mysticism, for example, you put yourself into that tradition and accept its basic rules. Progress usually takes place within these traditions – not across them. This was Feyerabend's main argument in his book *Against Method* (Feyerabend, 1975), which gained him a very questionable reputation as the guy who proclaimed 'anything goes'. In fact, what Feyerabend meant was the opposite. In art, just like in science, anything must not go. Knowledge or practical use can only be gained within and in relation to a specific tradition. The work of an artist or scientist is valued based upon this tradition. Whether all traditions are equally valuable in society ('anything goes') is a political, not a scientific question, argued Feyerabend.

In this respect, art and science are very much alike, much more, I think, than scientists and artists wish. Part of the appeal that the arts have for scientists is this misunderstood 'anything goes'. Many scientists envy artists for what they believe to be freedom from methodological constraints. It's true: artists do not quote fifteen pages of already existing literature only to be allowed to develop their own argument. But that is really only because they usually work in different media than theoretical text. Most artists I worked with sound extremely 'scientific' and like to quote a lot of literature when they write about their work.

On the other side, artists tend to idealize academic science for its rigour, morality and social authority. Morality and authority are the two trump cards in any scientist's deck. Artists tend to overlook how rigorous their own profession is and how strong the traditions that they build on. And as far as social authority is concerned, over the last decades academia has seen a severe deterioration of status and authority, which I will discuss below. This is often not acknowledged by artists, who have been themselves subject to scrutiny and even downright hostility throughout the history of modernity, particularly since the latest rise of political populism in Europe and the USA.

So what's the takeaway from this little excursion into Paul Feyerabend's 50-year-old theory? Two things: science and art are very alike. And: artists are not free. Still, freedom and innovation are obviously valued highly in our neo-liberal culture and in the arts as well. To illustrate freedom in the arts, I refer to one of the most famous artistic 'gestures', as US art critic Brian O'Doherty calls them in his history of the European Avant-garde (O'Doherty, 2000).

In 1917, the artist Marcel Duchamps submitted a urinal to the exhibition of the Society of Independent Artists in New York and called it 'The Fountain'. It was not exhibited but it was accepted, as, according to the statutes of the Society, any artwork by an artist was to be accepted, as long as the fee was paid. Duchamps' gesture was thus a gesture from an artist to his peers, fellow professionals, who, willingly or not, accepted it. Duchamps did not declare himself the author but he made clear his evaluation criteria: ready-made objects like 'The Fountain' were 'everyday objects raised to the dignity of a work of art by the artist's act of choice' (Martin, 1999). It's the artist's choice that creates value, not diplomas, technical mastery or the likeness or emotional appeal of the work. In other words: art is what the artist decides. Full stop.

But was this really new? The point is that Duchamps' stunt could only work because it referenced a social convention that was well established in Europe since the Romantic era. This period, the single-most important cultural revolution in modernity, essentially established the concept of the artistic authority. And by 1917 this concept had been well established through a whole set of (state-sanctioned) institutions, from the arts academy where artists taught other artists, to the national museums where artists decided which artists to be exhibited. It was a closed shop, as we would say today. Thus, Duchamps did not so much break with conventions as push the envelope. And it was not the public or any other outside force that shook the institution. In fact, only much later a photograph of 'The Fountain' found its way into public discourse.

What Audience?

Authority in the arts, like in science, traditionally is the result of an elaborate system of peer review. Usually, to get a spot in the spotlight, you have to know and be respected by a large number of professional peers. While some art forms require little material and institutional investment, becoming seen and heard does. Whether a visual artist, a musician or a choreographer, the audience can only love you when they know you. And while it is true that digital media today lowers the barriers significantly, it also increases competition drastically. And, at least as far as I can see, institutionalized cultural production has maintained the established hierarchies and authorities. When we admire artistic daring and radicalness in an artwork, this radicalness is usually performed in a tightly institutionalized setting, and very often without an audience. Does the audience then matter at all? Yes and no.

Another specificity of professional art is that it is not theoretical. Of course, the definition of theory and practice is unstable and historically shifting, but in general terms, I would argue that art works through manifestation. Art is an investigation of the terms, shapes, tones, colours and so forth, in which emotions or thoughts can be expressed. The investigation is carried out through experimentation, not theorization. Public exposure of course matters and is an integral part of artistic practice. However, much of what finds entry into the art canon had little to no public repercussions at the time of publication. The history of performance art in the 1960s and 70s is particularly insightful here (Van Mechelen, 2006). This is very important to realize and in fact is rather similar to academia.

To put it bluntly: art does not necessarily have an audience. Some artists don't care or are not even interested in having one. And it's not always a good idea to pick an artist for a collaboration based on their popularity. It depends on the goal of the collaboration. If artists and scientists decide to work together in order to find new or larger audiences, to popularize an idea or a scientific finding or to influence public opinion, things like TED Talks or science slams are a successful formula and you might not have to look much further. In other cases, however, designers, campaigners and journalists are much more suited for collaboration than artists.

Artists versus Institutions

In 2013 I developed a residency program for art-science collaborations in Stuttgart, Germany (Possmann, 2015). To set it up, I called a lot of university deans and heads of departments to make them enthusiastic about one-on-one collaborations between researchers and artists. I was very surprised to find that in German universities, institutions do not tell researchers what to do and that professors regard it as an insult if anybody tries to do so. This might have changed slightly over the last ten years, but the freedom of research remains an important cornerstone of academia – just like the 'freedom of art' (Schorlemer & Stoll, 2012).

In art-science collaborations, however, I think it's crucial whether and how institutions are included. The role of the cultural institution in shaping art is often overlooked. In the end, the directors, actors and painters are the stars of the show. But especially when it comes to public access and public awareness, it's the institutions that control it. The undervaluation of the institution is a particularly serious problem in art funding and cultural politics. Because here, like anywhere else, the attention goes to the stars, while it should go to the structures that enable the artists to work, and to work sustainably.

There are literally no employed artists, at least not in most of Europe and the USA. While we still have big classical orchestras with long-term employment contracts and unions, all choreographers, writers, directors, scenographers et cetera are freelance or are freelancing next to part-time employment. The same applies to visual artists, writers and filmmakers. Thus, the partnership between artist and institution is one of competition and negotiation; ideally, both partners are powerful enough to shape the outcome for the better.

Bypassing institutions in the arts is problematic, because it creates fragile and short-lived work relations and puts a lot of pressure and workload on the artist which would otherwise be shared by the institution. Making art is in most cases a very resource-intensive and costly business, despite what many people seem to think. Therefore, it is crucial to know what a specific cultural institution can supply and what it cannot. Asking an artist to create an artwork without including a cultural institution will result in endless debates about production budgets, spaces, technical equipment and so forth – all things of vital nature to any art work, even if it's just a painting.

It's interesting to note that also within the arts, the role of the institution is usually downplayed. Few artists openly admit that it is in fact the context that creates the aesthetic experience in an audience, not the artwork as such. But since contexts are notoriously fleeting, hard to control and even harder to analyze, our attention tends to go to the artwork as an object or quasi-object. This is for obvious reasons particularly problematic in the performative arts, where for decades the scientific discourse tended to focus almost exclusively on written text (drama) or scores (music or dance) rather than on what audiences actually witness and experience (Fischer-Lichte, 2008). But if art is all about applied communication, then place, time, atmosphere and people are what really matters (Mackintosh, 2023). And these are for the most part created and controlled by institutions. Clever artists know this and create strong and mutually beneficial working relationships with institutions. In fact, since there is no such thing as art without institutions, artists always work for and within specific settings. Very rarely have I encountered serious artists who just didn't care about the setting, whether it's a museum, a series, a festival, an addressee or technical media. Granting an artist 'total freedom' is actually more a sign of ignorance than of generosity and might even be regarded as an insult. For good reasons, I would say.

How to Be Popular

Many scientists take it to their hearts that the wider public finds them difficult to understand or their professional output inaccessible or elitist; too complicated to be meaningful. While I would argue that simple things often tend to be meaningless and not the other way around, the wish to be understood and thus to gain resonance (Rosa, 2019) is of course understandable. But unfortunately, artists by and large suffer the same pain and have no clue either how to solve this.

There are two structural reasons why art is not a popular business, at least in Europe and in the USA. Both are structural and not easily solved. One is the basic concept of development through critique, which the arts share with academic science. Professional artists conventionally relate to other artists and the tradition of art before them. And they usually are forced to do so within one genre only, as they are not able to master more. The longer and richer the tradition, the more advanced and the more specific their art will be. Eventually the craft becomes highly specialized and it becomes hard for any audience to follow these developments in a meaningful way. The other reason is that forms of communication are highly specific in the first place. Understanding art as I do, as a kind of applied science of communication, any understanding is necessarily limited to the confines of cultures and even subcultures.

Art is thus anything but universal. To make an analogy to science: if we describe artworks as experiments, then these experiments have very specific laboratory conditions. You cannot easily repeat an experiment in different conditions. The language barrier is the most obvious example, but I have witnessed countless times how dance pieces or performance shows have triumphed in one place and completely flopped in another. The same goes for all kinds of genres, even within one country or within the same city.

And sometimes time makes all the difference – what worked beautifully and excited audiences last year, might be irrelevant to the same audience today. It keeps astonishing me that the myth of universal art is maintained and how both artists and actors in other fields do not cease to call upon the arts to 'bring people together', 'to unite' and 'to bridge all differences'. Because if art is about culture, it is all about differences. Not the type of differences that lead to war and political oppression necessarily – although that does frequently happen – but the type of differentiation that makes social life fun in the first place.

Political Role – What is it For?

This of course brings us to the question of public or political legitimization: why do we need art or science? Why should it be publicly funded? I will not go into the various arguments for and against public funding, but rather compare the current environment that both professional groups find themselves in in Europe and the USA and what conclusions might be drawn from this.

If the arts are not universal, do not bring world peace and are not even relevant to 90% of the population in Germany – who needs them? (Almannritter, 2017). Obviously, this is a question that has troubled artists for a long time and particularly so during the years leading up to the Covid shut-down in 2020. From my observation, the main strategy in the European art scene to answer this scrutiny within the last decades has been to form alliances. One could say: the arts have tried to borrow legitimization or authority from other social fields, mainly from science and social work; sometimes more and sometimes less successfully.

In the professional environment I have been part of subgenres like conceptual dance and documentary theatre, which borrowed heavily from academia, and had been dominating the theatre stage at least since the 1990s. This movement from emotion to intellect, referred to in cultural studies as a discursive turn, is still quite relevant today. The discursive turn in the 1990s was followed by the social turn and a shift of attention from formal investigation to political and ethical discourse. While conceptual art and artistic research in the 1990s and 2000s tended to be conceptually challenging at best and self-centred, highly inaccessible and elitist at worst, social art was trying to bridge the class divide, create new access paths, tear down immaterial barriers and create alliances with non-art actors and fields like social work, campaigning, identity politics, and anti-capitalist, anti-racist and anti-colonial activism.

For cultural institutions like the publicly subsidized or run theatre and museum, this was not an easy task. In Germany, social workers, also largely dependent on public money, understandably saw their jobs and activities in competition and often in opposition to these cultural endeavours. Social work lacks the cultural capital and bourgeois appeal of the arts. Cultural work lacks street credibility and social capital among proletarian and marginalized communities. I often encountered animosity and rejection myself in trying to form alliances with migrant activists and social workers while working from inside the cultural institution.

With scientific institutions and professionals, this has proven somewhat easier. Scientific and cultural professionals by and large share the same bourgeois background and use and understand the same bourgeois codes as artists. This does not mean that either understand a lot about the other's profession at all. But at least they express their misunderstandings in the same language and have a similar level of self-confidence in dealing with them.

Yet I suspect there is something else at play. Art and science are both subject to critical scrutiny from various political directions. That's a pretty good reason to unite and, interestingly enough, both groups seem to think that the other is somehow more successful in fighting back. For several decades now, political populism across Europe has targeted artists and scientists, cultural and academic institutions; from the Dutch right-wing extremist Geert Wilders branding the arts a 'leftist hobby' (Fogteloo, 2010) in 2010 to Donald Trump's campaign manager Steve Bannon presenting 'alternative facts' and thereby discrediting scientific rationalism in 2017 (Blake, 2017). Both strategies proved extremely successful: the once exemplary performative arts system of the Netherlands is undermined, the environmental movement in the USA marginalized.

Despite these similarities, I see a profound difference in how populism is changing science and how it is affecting the arts. Due to the historical developments of art within capitalism since Romanticism, art now operates largely outside of the political arena, as a sort of alternative politics, alternative space and often enough as 'alternative facts' (Blake, 2017). And for a long time, artists have made themselves comfortable in that space and did not care whether they mattered all that much.

The social capital of scientists, however, is quite different. Remember the trump cards in any scientist's deck I mentioned before? This game is changing. By openly questioning the authority of scientific rationalism, populism is making use of and reinforcing a more general crisis of authority. It seems to me that populist strategists like Steve Bannon are in fact using the logic of the arts against science and other former authorities. Saying that anybody can have facts, and my facts are as good as your facts, is in fact like playing the artist card. If Marcel Duchamps can say 'an urinal is an art piece', so can I. Did Steve Bannon think of Duchamps' urinal, when he suggested in 2018 'the way to deal with them is to flood the zone with shit'? (Stelter, 2021).

Good Crisis, Bad Crisis

Like any good crisis, the crisis of authority is not simply good or bad. Whether you see this crisis of authority as a political disaster and profoundly unsettling, or whether you embrace it as a necessary step in social revolution, I guess is up to your authoritarian personality traits. Right-wing extremists did not bring it about; they are just using its effects for their cause and I don't see any reason why progressive socialists should not.

The question is: if our established 'knowledge traditions' (Feyerabend, 1975) had been all about argument and reason, why could a businessman like Donald Trump win major political office in 2016? I would argue that the 'alternative facts' claim worked, because by 2016 no one could honestly claim proof that scientific facts were any truer. The truth of the matter is that for most of us, scientists, journalists and politicians are only authorities as long as we agree to regard them as such. They are trusted because of a social contract that bestows authority regarding certain matters on certain professions, not because they are particularly convincing.

I close with what is to me another foundational moment in this crisis. In 2019, sixteen-year-old Greta Thunberg coined the rallying call 'unite behind the science' (Thunberg, 2019). The claim was successful for some time, but has by now disappeared completely from the public arena. This call for discursive order, if you will, was issued by no other than a very public figure that embodied the crisis of authority more than any other: the child that spoke truth to power.

The tale of Greta Thunberg can be seen as a modern-day myth and maybe the most lyrical political performance piece of the recent past: a child reminding the powerful to follow an established authority. Anthropologically speaking, the child or the mad person who defies authority and conventional wisdom is a profound and very old cultural figure. You might say that it was essential for the survival of social institutions of all kinds to question power in this way from time to time. So, in a time of profound crisis, Greta Thunberg became the ultimate authority on knowledge for the very reason that, as a minor, she could not be held accountable for delivering argument and proof.

That a child called on the UK Parliament to 'unite behind the science' is a piece of surrealist theatre as great as anything the surrealists ever wrote. Greta Thunberg's performance captures the same ambivalent desires I see in populist rhetoric or TED Talks: it's a simplistic and highly authoritarian gesture, offering a sense of clarity, reliability and security, paired with the promise that everyone is entitled to that stage and that role. We all can be Greta; we all can be the authority: a guide in times of information overload and disorientation, and a foot in the behind of the authorities. It's a desire as naive as a sixteen-year-old's sit-in in front of the national parliament.

References

Adorno, T. W., Horkheimer, M. & Kogon, E. (1989). *Die verwaltete Welt oder: Die Krise des Individuums. Ein Gespräch zwischen Theodor W. Adorno, Max Horkheimer und Eugen Kogon für den Hessischen Rundfunk, gesendet am 4. September 1950.* In: M. Horkheimer (ed.) *Gesammelte Schriften. Band 13: Nachgelassene Schriften 1949-1972*, 121-142 (Frankfurt am Main: Fischer).

Almannritter, V. (2017). *Audience Development in der Migrationsgesellschaft: Neue Strategien für Kulturinstitutionen* (Bielefeld: Transcript Verlag).

Blake, A. (2017). *Kellyanne Conway Says Donald Trump's Team Has 'Alternative Facts'. Which Pretty Much Says It All.* The Washington Post. Available: https://www.washingtonpost.com/news/the-fix/wp/2017/01/22/kellyanne-conway-says-donald-trumps-team-has-alternate-facts-which-pretty-much-says-it-all/ (accessed 18 June 2025).

Feyerabend, P. (1975). *Against Method: Outline of an Anarchistic Theory of Knowledge* (Berkeley: New Left Books).

Fischer-Lichte, E. (2008). *The Transformative Power of Performance: A New Aesthetics* (London: Routledge).

Fogteloo, M. (2010, 6 October). *Dag linkse hobby's.* De Groene Amsterdammer. Available: https://www.groene.nl/artikel/dag-linkse-hobby-s (accessed 18 June 2025).

Forschung & Lehre (2024, 16 May). *Studie zu Angriffen in der Wissenschaft veröffentlicht.* Available: https://www.forschung-und-lehre.de/zeitfragen/studie-zu-anfeindungen-in-der-wissenschaft-6423 (accessed 18 June 2025).

Glauner, M. (2019). *Get Involved! Participation as an Artistic Strategy.* Available: https://maxglauner.com/2019/07/11/get-involved-_-strategies-of-participation-in-contemporay-art/ (accessed 18 June 2025).

Kessler, S. (2011). *With 500 Million Views, TED Talks Provide Hope for Intelligent Internet Video.* Mashable. Available: https://mashable.com/archive/ted-anniversary (accessed 10 December 2024).

Mackintosh, I. (2023). *Theatre Spaces 1920-2020: Finding the Fun in Functionalism* (London: Bloomsbury).

Martin, T. (1999). *Essential Surrealists* (Bath: Parragon).

O'Doherty, B. (2000). *Inside the White Cube: The Ideology of the Gallery Space* (Berkeley: The Lapis Press).

Possmann, J.-Ph. (2015). *Dilettantische Foschun: Zehn Forschungsaufenthalte zwischen Kunst und Wissenschaft* (Berlin: Alexander Verlag).

Rosa, H. (2019). *Resonance: A Sociology of Our Relationship to the World* (Hoboken: Wiley).

Schorlemer, S., & Stoll, P.-T. (ed., 2012). *The UNESCO Convention on the Protection and Promotion of the Diversity of Cultural Expressions. Explanatory Notes* (Berlin: Springer).

Stelter, B. (2021, 16 November). *This Infamous Steve Bannon Quote Is Key to Understanding America's Crazy Politics.* CNN Business. Available: https://edition.cnn.com/2021/11/16/media/steve-bannon-reliable-sources/index.html (accessed 18 June 2025).

TED Organization. (2024). *About Our Organization.* Available: https://www.ted.com/about (accessed 10 December 2024).

Thunberg, G. (2019, 23 April). *"You Did Not Act in Time": Greta Thunberg's Full Speech to MPs.* The Guardian. Available: https://www.theguardian.com/environment/2019/apr/23/greta-thunberg-full-speech-to-mps-you-did-not-act-in-time (accessed 10 December 2024).

Van Mechelen, M. (2006). *De Appel: Performances, Installations, Video, Projects, 1975-83* (Amsterdam: De Appel).

Amnesia of a Forbidden Future

Kamar Ahmad Simon

The Bengal Delta, the world's largest river delta, stands at the confluence of colonial legacies, post-colonial geopolitics and mounting environmental crises. This paper explores the complex interplay of human agency and natural processes in shaping the region, with a focus on the socio-ecological impacts of river management on the Tista and Jamuna Rivers. Through an interdisciplinary lens, it revisits the colonial origins of water infrastructure, the enduring challenges of transboundary governance and the ecological consequences for marginalized communities in Bangladesh.

Drawing from personal experiences in creating a cinematic water trilogy, the paper interrogates how dominant narratives of modernity and development overshadow local knowledge and histories. It critiques the systemic inequalities embedded in global and national frameworks that continue to shape the delta's socio-environmental future. By integrating historical, geographical and cultural perspectives, this work highlights the urgent need for decolonized, inclusive approaches to address the region's challenges. The paper furthermore examines the role of cinema as a tool for amplifying marginalized voices, questioning whether it can transcend entrenched power structures and offer pathways for reclaiming subjectivities and narratives of resistance.

Once There Was a River

Like a ship with a broken mast, a small two-storied boat is stuck in the river bottom. The small wooden boat is abandoned near Balasi Ferighat, swaying in a sea of sand (Figure 1). Locals call the ship *M.V. Mohabbat Express*. Who knows where the last passengers went! Aptly named after their pump engines, such shallow boats are homemade vessels powered by modified diesel pump engines, originally designed for extracting shallow-surface water, primarily for farming. The initial M.V. of the vessel name is an acronym: M stands for Marine, V for Vehicle. 'Mohabbat' is a Persian word meaning 'love', while 'Express' obviously refers to speed. It must have been the fastest means of transport available, when there was a river. The word 'Mohabbat' echoes a time when Persian flowed through Bengal as its official language of court for over six centuries, from 1203 to 1837 (Bilah, 2021). 'M.V.' brings us back to the British occupation (1757-1947), which replaced Persian with English, reshaping the region's linguistics to serve the subjugation. Just like rivers have defined the delta's landscape, the history of Bangladesh is intertwined with its language and subtexts.

Balasi Ferighat, near Fulchori in the Gaibandha district of northern Bangladesh, is a stark reminder of the once mighty confluence of rivers like Tista and Brahmaputra. The Tista River, born in the Himalayas, winds 315 kilometres across the Indian states of Sikkim and West-Bengal, where it is heavily obstructed by more than two dozen dams. After this choking journey, the dying stream of the Tista flows another 109 kilometres into Bangladesh, where it meets the Brahmaputra, locally known as the Jamuna. The Brahmaputra River, as part of the Ganges-Brahmaputra-Meghna (GBM) river basin, is a vast transboundary system covering over 1.7 million km^2, with its area distributed among India (64%), China (18%), Nepal (9%), Bangladesh (7%) and Bhutan (3%). Nepal lies entirely within the Ganges River basin, while Bhutan is fully encompassed by the Brahmaputra basin. The Brahmaputra begins its journey in Tibet, winds through the Himalayas, and flows into Assam (India) before merging with Padma, the name of Bangladesh's part of the Ganges. The Ganges forms the largest river delta in the world, joining the Meghna and eventually reaching the Bay of Bengal. It has shaped the socio-cultural, economic and political lifelines of the region over millennia.

The course of the Brahmaputra and Meghna Rivers runs through mostly hilly terrain and it contains a lot of sand, which settles at the entrance of the plains of these two rivers. Therefore, the Bengal Delta is largely formed by sediments from the Ganges River, which are carried from the cultivated lands of the Ganges basin in central and northern India and deposited in the plains of Bengal. The Sundarbans, the estuary of this delta, was formed by the sedimentation of freshwater from upstream and vegetation growing in the estuary above (Chardi, 2024). Estimates indicate that Bangladesh had more than 1,200 rivers (Anam, 2024) even in the 1970s, intricately weaving through the nation as the last leg of the Himalayan river system. Today, however, only about 710 of these rivers are estimated to be still in existence, with many struggling to survive (Biswas, 2018). Among these, 57 are transboundary rivers, with 54 shared with India, highlighting the complex web of cross-border waterways (Alam, 2021).

The photo of the M.V. Mohabbat Express, stuck in the sand, was captured during one of my research trips in northern North Bengal, on the Bangladesh side. I have frequented the region since 2016 in search of the final script of my Water Trilogy. This photo echoes the opening scene of *Are You Listening!* and the closing of *Day After . . .*, the first two parts of my trilogy. The story of this trilogy began back in 2009, when I stumbled upon a picture in the daily newspaper. The picture captured thousands of people, armed with only spades, desperately trying to hold back the water from a broken *bund* by piling up clay mounds with their bare hands. *Bunds* and *bundals* are simple earthen walls, strengthened with bamboo, of an indigenous design learned through generations, which safeguard farmland and settlements from river erosion and seasonal floods. The *bund* in the newspaper was broken after a tropical cyclone named Aila swept through and ravaged the coasts of Bangladesh. In 11 out of 64 districts, the cyclone caused widespread devastation, damaging or destroying around 600,000 thatched homes, 8,800 km of roads, 1,000 km of embankments and 123,000 hectares of land. The disaster impacted approximately 9.3 million people, leaving 1 million homeless (Walton-Ellery, 2009).

The Bengal Delta has long relied on natural flooding cycles to sustain its agrarian traditions. As a result, *bunds*, *bundals* and larger dams have been central to water resource engineering in the Indian subcontinent, designed to protect farmland and settlements from seasonal floods. These low-lying earthworks reflect the region's deep-rooted expertise in managing water flow. One of the earliest examples of flood control infrastructure is the Kallanai Dam, built in 150 CE by Chola king Karikala in Tamil Nadu. Similarly, in the twelfth century, Kaivartya king Bhima constructed the Bhimer Jangal (Khan, 2021), a highway-cum-embankment still known in northern districts like Bogra, Sherpur and Gaibandha, aimed at protecting the area from annual monsoon floods. However, in recent decades, the frequency and intensity of cyclones in the south have tripled, while in northern Bangladesh, most flash floods are caused by the release of water from large concrete dams built across almost all transboundary rivers with India.

Purus of the Delta

Something in the newspaper picture reminded me of the works of SM Sultan (1923-1994). Born in Narail, in southern Bangladesh, Sultan became one of the leading decolonial painters from Bangladesh. He explored the world and, although he did not complete his studies at Calcutta Art College, he took part in an exhibition in London with artists like Dali, Matisse, Picasso, John Brack and Paul Klee in the early fifties. Returning to his village in Narail by the River Chitra (an offspring of the Padma), Sultan embarked on creating evocative paintings employing jute canvas and locally sourced pigments made of herbs, vegetables and plants (Figure 3). Through his masterful brushstrokes, Sultan adeptly captured the villagers' indomitable spirit and resilience in the face of the myriad challenges posed in the ordinary lives in the delta. The peasants in his canvases resemble the bison hunters of Altamira's cave paintings – children of the untamed soil, bounded by a timeless vow to cultivate the land.

The positioning of the delta in Sultan's canvas has profoundly influenced me since my student days. I visited Mehendiganj, the village shown in the newspaper picture, nestled in the estuary where the Meghna River meets the Bay of Bengal. Captivated by the stories of fishermen and peasants, for the next three months, I journeyed 200 km along the riverways towards the east, up to the Indian border. The line was drawn in 1947 by the British lawyer Cyril John Radcliffe, who had little understanding either of the delta or the hydrography of the Himalayan river basins (Pillalamarri, 2017). My trips have taken me to dimly lit tea shops, filled with thick wood smoke. Characters like those in Sultan's canvases have shared their stories of Sundarbans, of the mangrove forest, of the Bengal tiger and of Puru, the ghost-spirits of those who got lost in the woods and never returned. Their pirate tales recounted the Portuguese, who turned the Bay of Bengal and the Sundarbans into their hunting grounds, capturing men and women to sell as slaves across the globe. Accounts of people displaced by rising waters sounded like a never-ending saga of the delta, which has a past but no future.

For three years, I lived among villagers in southern Bangladesh who had rebuilt their lives atop a *bund* after coastal surges swept their land away. This journey became *Are You Listening!* (Simon, 2012), a 90-minute non-fiction feature film exploring deception, betrayal and the politics of those on the margins in the southern delta. Yet, their story felt incomplete. Characters like Puru haunted me, pushing me to connect missing links – land loss in northern Bangladesh, cross-border river dynamics and the erosion of livelihoods under postcolonial statehood. Compelled by their voices, I embarked on a loosely-connected water trilogy. From the southern delta to the capital, I travelled aboard a century-old paddle steamboat, a relic of British colonialism, which became both a character and a stage. It carried a diverse cast – rich and poor, powerless villagers and powerful politicians – all caught in a post-colonial conundrum. This eight-year journey culminated in *Day After* . . . (Simon, 2021), a 115-minute hybrid feature capturing these vivid contrasts and everyday struggles. Yet, the film struggled to fully link the rivers' troubled legacy with the socio-political crises of the modern delta.

For *Till the Last Drop;* , the final chapter of my water trilogy, I turned to northern Bangladesh, where rivers are dying. Since 2016, my focus has been on the intricate Tista-Brahmaputra confluence and its vanishing islands, anchored around Balashi Ferighat in Fulchori, Gaibandha. A metafiction unfolds here, led by a 50-year-old school teacher, Faruq Hossen, and his troupe of misfits. They recreate scenes intertwining the historic fall of Sirajuddaula to the British East India Company in 1757 and the recent uprising that toppled Hasina's regime, long seen as India's puppet. Lokman and Laskar, two brothers, roam the barren sands with their buffalo, searching for remnants of the once-mighty rivers. Hazrat, a passionate artisan, works with his team near Balashi Ferighat to construct their own Noah's Ark, bracing for the inevitable floods when upstream dams are released during the monsoon.

Between History and Memory

Around 325 BCE, Greek and Latin historians wrote that Alexander withdrew from his Indian campaign to avoid confronting the powerful Gangaridai, a kingdom named 'Ganges in the heart', centred on the delta (Chowdhury, 2021). Spanning 350 km, the delta nurtured civilizations as ancient as the Nile, Indus and Euphrates, yet distinct in its philosophy, social structures and agriculture. Known for its six-season agricultural cycle and three-crop calendar – *aush*, *aman* and *rabi* – it supported flourishing empires like the Mauryan, Gupta, Pala and Sena between the third century BCE and the twelfth century CE (Goldberg, 2024a, b; Zeidan, 2024, 2025). The Sompur Bihar ruins, the second-oldest residential university after Nalanda, attest to the delta's Buddhist scholarly legacy. Evidence suggests Persian Sufis arrived in the eighth century, providing refuge to oppressed lower-caste and indigenous peoples, blending local traditions to create a unique Bengali Islam rooted in resistance. Between the fourteenth and sixteenth centuries, Bengal's independent sultanate defied Delhi until the Mughals took control. By 1700, Mughal India was the world's largest economy, with Bengal as its wealthiest province. Renowned for textiles, ship building and high living standards, Bengal supplied 40% of Dutch Asian imports (Hancock, 2022), solidifying its place as a global economic hub.

The British East India Company's control over the Indian subcontinent began with the Battle of Plassey in 1757, where political manoeuvring and alliances with local factions led to the defeat of Bengal's ruler (Roychowdhury, 2024). This event marked a significant shift in regional governance and economic systems, allowing the Company to consolidate power through social oppression and resource extraction. In 1765, the Company secured Mughal-sanctioned Diwani rights to collect revenue in Bengal, Bihar and Orissa. The subsequent implementation of stringent taxation policies, coupled with environmental factors, severely impacted the agrarian economy. These conditions contributed to the 1770 Bengal famine, which resulted in the deaths of an estimated 10-15 million people, illustrating the profound socioeconomic and demographic consequences of colonial administration.

Famine became a recurring tragedy during two centuries of British occupation, culminating in the Bengal Famine of the 1940s, which killed five million people. Beyond the famines, British occupation reshaped Bengal's landscape and livelihoods, leaving irreversible damage. Before British intervention, Bengal's intricate network of over a thousand rivers sustained millions through natural flooding cycles and river-based agriculture. Villages were interconnected via canals, linking sub-districts and districts through major rivers with a significant riverway every 15 km. Travel spanned up to 120 km during the wet season, employing millions in river transport (Biswas, 2018). The *ganj*, river entrepôts, served as critical commercial hubs, connecting micro- and macro-economies. However, under the guise of development, colonial engineering projects, such as railway bridges for faster military mobilization and dams to expand taxable farmland (Gilmartin, 1994), disrupted this millennia-old ecosystem. The rise of markets near railway stations gradually replaced the *ganj*, undermining the socio-economic symbiosis of the riverine trading and agriculture that had thrived for centuries.

The British occupation of Bengal, lasting longer than in any other part of the Indian subcontinent, faced persistent rebellions of the indigenous people and the peasants, from the Chuar Rebellion (1766-1834) to Fakir-Sannyasi Uprising (1771), Titumir's Narkelberia Jang (1831), the Santal Rebellion (1855) to the First Indian War of Independence (1857-1859), described by Karl Marx and Friedrich Engels (1959) as a significant resistance. Despite the resistance, by the 1860s, the British replaced the East India Company with the Imperial Civil Service (ICS) to promote the so-called 'Bengal Renaissance'. By the 1872 census, the British fostered a discrimination policy categorizing people by religion, in opposition to the broader Indian concept of Dharma (Baghat, 2006). This solidified the Hindu-Muslim divide, enabling the British to maintain control over the next century. By the time of India's partition in 1947, this divide had become irreversible. Bengal was split into Hindu-majority West Bengal, part of India, and Muslim-majority East Bengal (later East Pakistan), part of Pakistan, with a 2,204-kilometer border separating the eastern from its western counterpart. Post-independence, colonial legacies and communal divisions continued to shape the region.

During the two centuries of British occupation, along with selectively constructed histories, much of the region's indigenous science and technology was erased through violence. The colonial ideas of 'liberalism', 'modernity' and 'enlightenment' brutally disrupted thousand-year-old local skills and knowledge in medicine, engineering and architecture. The colonial myth of development, grounded on Eurocentric models, was ingrained in the education system so deep that even after independence, it continued to rule the worldview of the political class of India and Pakistan and shaped their vision for the future. Both the newly liberated nations neglected the intricate ecological balance built over millennia of their common hydro-basins, unilaterally building an infrastructure of irrigation and flood control projects. The lack of a common vision on the river basin not only intensified the region's environmental degradation but also disrupted traditional water management systems passed over generations for centuries.

Following the military crackdown in East Pakistan in 1971, characterized by widespread violence and significant loss of life, India intervened militarily in support of East Bengal. The nine-month conflict, referred to as the War of Independence in Bangladesh and the India-Pakistan War in other contexts, culminated in the emergence of Bangladesh as an independent, Muslim-majority state. The new nation inherited 57 transboundary rivers, along with the environmental challenges associated with shared water resources. India's upstream infrastructure development, including dam construction along key rivers, has altered the natural flow in the Bengal Delta, impacting ecological systems and resource availability. These changes have disproportionately benefited India's West Bengal region while creating ecological and socioeconomic challenges for Bangladesh. Disputes over equitable water sharing have persisted, with Bangladesh often perceiving India's policies as dismissive of its concerns. As the downstream estuary of the Ganges-Brahmaputra-Meghna basin, Bangladesh faces significant vulnerabilities, with up to 67% of its land submerged during monsoons and only 11% during the dry season (Bangladesh Planning Commission, 2018). Water diversions via dams such as the Farakka, Tista and the proposed Tipaimukh Dam exacerbate river erosion and siltation, disproportionately affecting Bangladesh due to its geographical position and dependence on these shared water systems.

Shortly after independence in 1947, India's Prime Minister

Jawaharlal Nehru hailed dams as the 'temples of modern India', launching projects like the Farakka Barrage, first proposed in 1853 by British engineer Arthur Cotton to clear silt from Kolkata's port. Despite warnings from engineer Kapil Bhattacharya about severe flooding and siltation in East Bengal (Bangladesh) and West Bengal, the dam opened in 1975, causing the Padma River in Bangladesh to dry up, triggering desertification and ecological collapse. In 2017, Bihar's Chief Minister Nitish Kumar called for the dam's decommissioning, blaming it for floods in Bihar (Chari, 2016). The Tista River, flowing through Sikkim, West-Bengal, and a major tributary of the Jamuna (the downstream Brahmaputra in Bangladesh), reflects another post-colonial dispute. Proposed for irrigation by the British in 1935 (Sarker et al., 2011) and revived in the 1960s, both India and Bangladesh built similar dams just 100 km apart in the late 1980s, exacerbating ecological crises. India now operates 28 hydroelectric projects on the Tista, generating 5,494 MW but devastating the river's flow and causing siltation across northern Bangladesh. The Tipaimukh Dam, first proposed by the British in 1926 on a tributary of the Meghna, is now under construction near Bangladesh's northeast border, intensifying hydropolitical tensions and threatening the livelihoods of the delta's once-thriving communities (Thakur, 2020).

Shadows of the West

In *Day After . . .*, the second part of my water trilogy, a native traveller asks a German tourist how to reach the 'Europe country'. For the tourist, the river journey is an exotic adventure, but for the native, the German represents Europe as a dreamland with solutions to his suffering – an image rooted in colonial-era reverence. This post-colonial view extended from the fine arts to water resource engineering. By 1957, the British Minister Julius Krug proposed a flood control plan (Faruque, 2021). Drafted in 1964 by American firm IECO, it led to embanked polders that initially benefited agriculture but later disrupted natural silt deposits, worsening flooding and waterlogging in the delta (Gani Adnan et al., 2019). After the 1947 independence, the 1950s-60s saw artists from Bangladeshi (then East Pakistan) studying in Europe, while decolonial artist SM Sultan rejected this and fled London. By the 70s-80s, as Edward Said (1978) critiqued 'orientalism', many settled in Europe, occasionally returning to extol Europe as the knowledge epicentre. Institutions like the British Council, USIS Library, Goethe-Institut and Alliance Française reinforced this pro-Western mindset. The US Agency for International Development, in an effort to counter the spread of the ongoing 'Red Revolution' in Asia, including the Naxalite-Maoist uprisings in Bengal, facilitated the export of the 'Green Revolution'. During the pro-democratic movements of the 80s-90s, as Dipesh Chakrabarty (2000) wrote in *Provincializing Europe*, intellectuals – echoing Frantz Fanon's (1952) critique of 'white masks' – shaped Bangladesh's cultural landscape with a westernized lens (Schmalzer, 2015).

Are You Listening!, the first part of the trilogy, features villagers constructing *bundals*: simple yet effective flood control structures passed down through generations, using local resources and collective effort. This tradition starkly contrasts with neo-liberal capital-driven models and challenges the alienation Marx warned about, especially in matters of survival. In contrast, the Bangladesh Delta Plan 2100, devised under the recently toppled Hasina regime – which was marked by corruption, violence and kleptocracy – ignored traditional wisdom and deltaic complexities (Khalequzzaman, 2016). Developed by Dutch consultants with knowledge of the regime's exploitative practices, the plan reflects earlier ecological challenges, such as NEDECO's 1960s polder projects in East Pakistan (now Bangladesh). These embankments prevented seasonal floods and supported agriculture at the cost of natural silt deposition cycles, causing waterlogging and salinity. By the 1970s, these issues had worsened, altering the hydromorphology of southern Bangladesh and severing river-floodplain connectivity in the north. As downstream water flow decreased in the Padma River due to the Farakka and Tista Dam projects on the other side of the border, the upstream tidal pressure increased, raising the riverbeds in the Sundarbans and worsening the salinity levels. Over time, this has exacerbated environmental degradation in southern Bangladesh, particularly in Satkhira and Khulna, where alluvial deposits from the Ganges no longer replenish the land. As a result, these southern regions are sinking, while northern Bangladesh experiences a natural rise of about 3 millimetres per year due to alluvial deposition. Although NEDECO's interventions initially provided protection, they highlight the dangers of large-scale, poorly integrated flood control projects (Islam, 2016).

Cinema as an Iceberg

To me, cinema often resembles an iceberg. What appears on screen is merely the tip of a much larger and deeper reservoir of emotions, dilemmas and scars. My journey has traversed a wide spectrum of experiences, from facing local challenges to get my films released, to receiving prestigious invitations from renowned festivals such as IDFA, Berlinale, Locarno and Cannes, among others. While gathering field notes for the first chapter of my trilogy, *Are You Listening*! (2012), I traversed nearly the entire coastal belt of southwest Bangladesh. Since 2016, I have been drifting along the sandy islands where the Jamuna River converges with the desolate Tista in the north, compiling notes for the final film of the trilogy, *Till the Last Drop;* . The Covid hiatus provided a necessary pause, allowing me to reflect on the journey I have undertaken since 2010. I was struck by the depth of pain and untold stories that often remain buried beneath the surface during the filmmaking process. The self-censorship I encountered came not only from local authorities but also from international bodies adhering to modernist ideals of objectivity and universalism. Their cultural biases, influenced by coercive frameworks and patronization practices, prioritizing 'important others' over a more fragmented and pluralistic view of reality that includes 'lesser others', profoundly impacting my thoughts and writing.

Despite receiving critical acclaim, *Are You Listening!* faced significant delays, remaining shelved for months while awaiting approval from the Bangladesh Film Censor Board. Established under the British-era Cinematography Act of 1918, the board has evolved through various political regimes, with changes in name and amendments, yet its role in regulating content, including potential suppression of dissent, remains a point of contention.

In the case of *Day After...* (2021), the film faced prolonged challenges, with the Certification Board (formerly the Censor Board) withholding mandatory clearance for over a year. The certificate was finally issued in March 2025, largely thanks to the July Uprising of 2024, though it still required considerable effort to get the film cleared. An official response from the board raised concerns over its themes, suggesting possible political sensitivities. The project, produced under Arte-France's prestigious La Lucarne program, also encountered creative disagreements. The Arte commissioner – a European writer, producer and curator – disagreed with the filmmaker's vision for the final cut, with the French producer advocating for a shorter version. These creative differences escalated when the Norwegian producer supported the French perspective, highlighting complex power dynamics within the collaboration.

This experience illustrates the challenges faced by independent filmmakers from the Global South when working within European funding structures. It underscores how creative disagreements can be compounded by imbalanced power relations, adding layers of difficulty to the production process. A simple email exchange between the French and Bangladeshi teams quickly escalated into a diplomatic issue when the French producer complained to the French consulate in Bangladesh, which forwarded it to the Ministry of Foreign Affairs. The conflict was eventually resolved through intervention of the former Director of the International Documentary Film Festival Amsterdam-Bertha Fund Director, but this involvement came only after the film was selected for IDFA's main competition. This underscores that the intervention was aimed to support the filmmaker as a co-producer, as well as to ensure the project was salvaged for the festival, reflecting the often-subjective nature of international support in such productions. As I prepare to wrap the final chapter of the water trilogy, I'm daunted by the task of interpreting this baggage of the past, given the question of selective freedom today. As dams choke my delta, reshaping my landscape, I struggle to craft the story of a 'lesser other' to compete with the 'urgent' and 'important' others. My burden is shaped by my limited choices and my relevance within this transboundary captivity, century-old colonial-politics hovering over me in the name of unfolding politics.

Questioning the 'Question Itself'

In a post-global world, when the international mainstream is fed as a local trend, even distant conflicts like the war in Ukraine become subjects of conversation at the family dining table in Dhaka. Meanwhile, critical local issues, such as the challenges surrounding the Tista River, often fail to garner sufficient attention within their own context. Furthermore, the ability to discern the truth is increasingly challenging, as media outlets often present events through the lens of their own biases, shaping public perception and discourse. Narratives of denial and oversimplification often dominate, obscuring deeper complexities. Similarly, the pressing global crisis of climate change transcends its surface-level framing, revealing underlying political and systemic failures that remain unresolved beyond seminar discussions and global events like COP conferences.

The prioritization of military expenditures over climate action exemplifies this disconnect. In 2021, global military spending reached $2 trillion, while only 5% of that amount – $100 billion – would have been sufficient to significantly mitigate climate risks. This reflects a broader trend of development strategies that marginalize large portions of the global population, leaving them excluded from dominant discourses and disproportionately burdened by their consequences. In this era of post-truth, questioning foundational assumptions has become increasingly difficult, as logical fallacies and denial often take precedence over critical reasoning and meaningful action. To question the 'question' itself appears impossible.

When climate risk management funds are converted into loans, trapping climate-vulnerable countries in a new debt cycle (Woolfenden & Khushal, 2022; Oxfam, 2023), the western narrative of climate justice loses its relevance in the global south. In my opinion, 'climate change' now rather appears like a distraction, obscuring the ongoing impacts of colonization in the new names and forms, much as 'human rights' have been weaponized against the south instead of addressing the question of injustice by the north.

As I have witnessed, the post-Covid period pushed for de-globalization instead of rights to vaccines for all. Climate issues are likely to be sidelined, if not forever, at least until a new global order is established in this evolving multi-polar world. Almost half (44%) of the world's habitable land is used for food production; two out of every three people are likely to be living in cities by 2050 in a world with a population near about 10 billion. My hypothesis is that cinema, when used as a tool of hegemonic power to assert selective objectivity and preferred agency in dominating metanarratives, fails to trace the colonial roots of my scripts, preventing me from raising the question of subjectivity. This adds to my dilemma, as I realize more and more that cinema is being used as a 'global integration' project for the south, and the old-spirit of the never-ending enterprise of 'civilizing' or 'modernizing' by the north endures.

As a filmmaker, navigating national grants requires aligning with dominant local political narratives, while on the international stage, the pressure shifts to conforming to selective narratives favoured by western funders. These alignments often reproduce the grand narrative of the west, leaving little room for alternate streams. In today's polarized world, binary choices erase middle grounds and block grey zones, leaving devastating impacts. Every year, 50,000 of my compatriots are displaced due to flooding and erosion, and with climate change accelerating, rising sea levels could render 20 million people homeless by 2050 (UNB, 2021). Yet, such urgency is overshadowed by reproductive 'mainstream' narratives. The key question remains: how can a filmmaker from the global south remain politically correct while challenging these interconnected frameworks that often seem impossible to confront?

The Role of Cinema

Libraries close, fleeting images dominate learning, corporate social media overtakes traditional broadcasting and AI algorithms are shaping our decisions. Is this a risk or an opportunity? When television came to Bangladesh (then East Pakistan), it was hailed as a revolutionary tool for 'distance education', but devolved into a medium for trivial entertainment, altering culture within the confines of urban life. Satellite channels brought promises of globalization yet ultimately colonized local culture. The internet, once seen as a platform for global participation, amplified existing power structures, leaving many as passive consumers rather than active participants. Each technological advance carries the risk of control unless critically reclaimed. As AI has started to write film scripts, how do we ensure it doesn't consume the selective story of 'them' and render 'us' deaf or dumb subjects again? Though cinema wields unmatched potential, its purpose often eludes even its creators. Every morning, as I draft papers and pitches for *Till the Last Drop;* , the final chapter of my Water Trilogy, doubts consume me. What is the point of dedicating my best years to a trilogy if it neither preserves the history of my water nor confronts the existential challenges facing my delta?

As an architect-turned-filmmaker, it took me a decade to trade my pen for a camera, followed by another decade of unlearning to find my own cinematic language. My heritage – thousands of years of memory carried in my genealogy, centuries of history shaped through ontology and epistemology, and a decade of festival experiences – shapes my choices and leaves me questioning: where do I take the next ten years, perhaps my last decisive years, without failing to uncover the deeper truths beyond the subtexts? Cinema, as the youngest and most dynamic art form, connects billions instantly, yet remains tethered to mechanisms inherited from colonial enlightenment enterprises. Narratives become doctrines, and festivals celebrate palatable stories while sidelining 'lesser' others as mere sidebars. Despite its transformative potential, cinema often fails to recognize and internalize the wounds of history, leaving it burdened by centuries of unresolved grief and guilt. In an age of digital reproduction, to borrow from Walter Benjamin (2008), cinema still falls short of healing collective pain or addressing deeper subjectivities, raising the question of whether it can ever fulfil its profound promise.

Figure 1: Balashi Ferighat, Gaibandha, northern Bangladesh, 2022. Photo: Kamar Ahmad Simon.

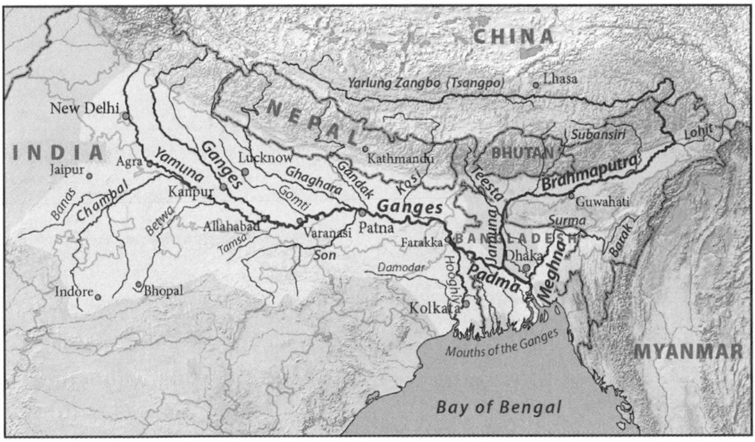

Figure 2: Ganges-Brahmaputra-Meghna (GBM) basin. Pfly, Wikimedia Commons, 2011.

Figure 3: Journey (যাত্রা). Oil painting, 120 x 180 cm. S.M. Sultan, 1986.

Figure 4: Still from *Are You Listening!*(শুনতে কি পাও). Kamar Ahmad Simon, 2012.

References

Alam, M. S. (2021). *River and Drainage System*. Banglapedia. Available: https://en.banglapedia.org/index.php/River_and_Drainage_System (accessed 15 January 2025).

Anam, M. (2024). *Of Rivers and People: How We Neglect Our Two Biggest Assets.* The Daily Star. Available: https://www.thedailystar.net/opinion/views/the-third-view/news/rivers-and-people-3741766 (accessed 15 January 2025).

Baghat, R. B. (2006). 'Census and Caste Enumeration: British Legacy and Contemporary Practice in India', *Genus 62*(2), 119-134.

Bangladesh Planning Commission (2018). *Bangladesh Delta Plan 2100.* Ministry of Planning, Government of the People's Republic of Bangladesh.

Benjamin, W. (2008) [1935]. *The Work of Art in the Age of Mechanical Reproduction* (London: Penguin Books).

Bilah, A. (2021). *Persian.* Banglapedia. Available: https://en.banglapedia.org/index.php?title=Persian (accessed 15 January 2025).

Biswas, A. (2018). *Rivers of Bangladesh*. [original language Bengali].

Chakrabarty, D. (2000). *Provincializing Europe*. (Princeton: Princeton University Press).

Chardi, S. (2024). *The Geological Formation of the Sundarbans: A Marvel of Nature's Engineering.* Greater Sundarbans. Available: https://greatersundarbans.org/blog/geological-formation (accessed 15 January 2025).

Chari, M. (2016). *Over 50 Years Ago, Bengal's Chief Engineer Predicted that the Farakka Dam Would Flood Bihar.* Scrool.in. Available: https://scroll.in/article/815066/over-50-years-ago-bengals-chief-engineer-predicted-that-the-farakka-dam-would-flood-bihar (accessed 15 January 2025).

Chowdhury, A. M. (2021). *Gangaridai.* Banglapedia. Available: https://en.banglapedia.org/index.php/Gangaridai (accessed 15 January 2025).

Fanon, F. (1952). *Peau noire, masques blancs*. (Paris: Éditions du Seuil).

Faruque, H. S. M. (2021). *Water Resources Management.* Banglapedia. Available: https://en.banglapedia.org/index.php/Water_Resources_Management (accessed 15 January 2025).

Gani Adnan, M. S., Haque, A., & Hall J. W. (2019). 'Have Coastal Embankments Reduced Flooding in Bangladesh?' *Science of The Total Environment 682*, 405-416. https://doi.org/10.1016/j.scitotenv.2019.05.048

Gilmartin, D. (1994). 'Scientific Empire and Imperial Science: Colonialism and Irrigation Technology in the Indus Basin', *Journal of Asian Studies 53*(4), 1127-1149. https://doi.org/10.2307/2059236

Goldberg, M. (2024a). *Pala Dynasty*. Britannica. Available: https://www.britannica.com/topic/Pala-dynasty (accessed 15 January 2025).

Goldberg, M. (2024b). *Sena Dynasty*. Britannica. Available: https://www.britannica.com/topic/Sena-dynasty (accessed 15 January 2025).

Hancock, J. (2022). *The English and Dutch East India Companies' Invasions of India*. World History Encyclopedia. Available: https://www.worldhistory.org/article/2048/the-english-and-dutch-east-india-companies-invasio/ (accessed 15 January 2025).

Khalequzzaman, M. (2016). *Bangladesh Delta Plan 2100: Not the most practical proposal*. The Daily Star. Available: https://www.thedailystar.net/op-ed/politics/not-the-most-practical-proposal-1211857 (accessed 15 January 2025).

Khan, M. A. (2021). *Bhimer Jangal*. Banglapedia. Available: https://en.banglapedia.org/index.php/Bhimer_Jangal (accessed 15 January 2025).

Marx, K., & Engels, F. (1959) [1857-1859]. *The First Indian War of Independence* (Moscow: Progress Publishers).

Oxfam (2023). *Rich Countries' Continued Failure to Honor Their $100 Billon Climate Finance Promise Threatens Negotiations and Undermines Climate Action*. Oxfam International. Available: https://www.oxfam.org/en/press-releases/rich-countries-continued-failure-honor-their-100-billon-climate-finance-promise (accessed 15 January 2025).

Pillalamarri, A. (2017). *70 Years of the Radcliffe Line: Understanding the Story of Indian Partition.* The Diplomat. Available: https://thediplomat.com/2017/08/70-years-of-the-radcliffe-line-understanding-the-story-of-indian-partition/ (accessed 15 January 2025).

Said, E. (1978). *Orientalism*. (New York: Pantheon Books).

Sarker, D., Zerin, A. I., Pramanik, B. K., & Ara, I. (2011). 'Climatic Impact Assessment: A Case Study of Teesta Barrage Irrigation Project in Bangladesh', *International Journal of Civil & Environmental Engineering 11*(1), 75-81.

Schmalzer, S. (2015). *Red Revolution, Green Revolution: Scientific Farming in Socialist China* (Chicago, IL: University of Chicago Press).

Simon, K. (director, 2012). *Are You Listening!* [film]. Sara Afreen for Beginning Production Ltd.

Simon, K. (director, 2021). *Day After . . .* [film]. Sara Afreen for Beginning Production Ltd.

Thakur, J. (2020). *India-Bangladesh Trans-Boundary River Management: Understanding the Tipaimukh Dam Controversy*. Observer Research Foundation. Available: https://www.orfonline.org/research/india-bangladesh-trans-boundary-river-management-understanding-the-tipaimukh-dam-controversy-60419 (accessed 15 January 2025).

UNB (2021). *Rising Sea Levels Can Render 20m Bangladeshis Homeless by 2050*. The Daily Star. Available: https://www.thedailystar.net/environment/climate-crisis/climate-loss/news/rising-sea-levels-can-render-20-million-bangladeshis-homeless-2050-2175646 (accessed 15 January 2025).

Walton-Ellery, S. (2009). *A Review of the Cyclone Aila Response 2009.* Review of the IFRC-led Emergency Shelter Coordination Group. Dhaka. Available: https://go-api.ifrc.org/publicfile/download?-path=/docs/evaluations/Evaluations2010/Asia%20Pacific/Bangladesh/&name=Review_of_cyclone_Aila_Response__22%2003%20 10SWE.pdf (accessed 15 January 2025).

Woolfenden, T., & Khushal, S. S. (2022). *The Debt and Climate Crises: Why Climate Justice Must Include Debt Justice.* Debt Justice. Available: https://debtjustice.org.uk/wp-content/uploads/2022/10/Debt-and-the-Climate-Crisis-Briefing-October-2022-UPDATED.pdf (accessed 15 January 2025).

Zeidan, A. (2024). *Mauryan Empire*. Britannica. Available: https://www.britannica.com/place/Mauryan-Empire (accessed 15 January 2025).

Zeidan, A. (2025). *Gupta Dynasty*. Britannica. Available: https://www.britannica.com/topic/Gupta-dynasty (accessed 15 January 2025).

Herwinnen door werken, Alex Roosdorp & Marie Roosdorp-van den Berg, 1945 + *Walcheren*, Jo de Haas & Mannus Franken, 1945

Rommy Albers

Alex Roosdorp made a large number of agricultural public information films in the 1930s and 1940s, including on the grain harvest in the new Wieringermeer polder and on the reclamation of the Northeast Polder. In April 1945, after the liberation of his hometown Deventer by the Allies, he began filming the devastated Netherlands. During the summer months of 1945, he and his wife Marie Roosdorp-van den Berg travelled the country by bicycle, filming the battered and destroyed landscape and infrastructure. The final result was an eighty-minute documentary, *Herwinnen door werken* (*Reclaiming by Works*), in which they show not only the devastation but also the first attempts at recovery and reconstruction.

The Roosdorps also visited two flooded areas on their trip: Walcheren and the Wieringermeer. On Walcheren, the sea dikes near Westkapelle and Vlissingen among others had been bombed by the Allies on 3 October 1944. This flooded much of the island and killed some 180 people in Westkapelle. A month later, Walcheren was occupied by Allied troops and the sea route via the Western Scheldt to the port of Antwerp fell into Allied hands. The Wieringermeer was destroyed by retreating German troops in April 1945. They blew up the dike in two places, after which the polder flooded in two days. In *Herwinnen door werken*, the results of these actions are shown: the land of Walcheren, made infertile by being flooded by saltwater, and the apocalyptic Wieringermeer, where the remnants of buildings stick out like palisades above the surface of the water.

It was not only the Roosdorps who filmed on Walcheren and in the Wieringermeer in 1945. The filmmakers Jo de Haas and Mannus Franken also made films about the destruction and reconstruction of both Walcheren and the Wieringermeer. In these, the new future is also shown to emerge. Because the infrastructure in Walcheren had disappeared, land consolidation allowed for more modern and larger-scale farms. Some of the farmers for whom there was no room left were given priority to move to the newly reclaimed and designed Northeast Polder.

Liquid Cinematic Spaces

Nina Bačun

This paper explores the fluid and dynamic nature of cinematic space within the contexts of climate change, digital technology, and modernity. Drawing on Zygmunt Bauman's concept of 'liquid modernity,' it examines how film reflects the instability of contemporary society, contrasting abstract spaces with meaningful places shaped by human experience. Emphasis is placed on the role of cinematic space in framing identity and subjective experience, with particular attention to how film techniques such as montage and camerawork contribute to this perception. Cinematic space is conceptualized as an active counter-archive – one that preserves and reinterprets cultural memory in the digital era.

The discussion addresses the evolving nature of film as both a 'living' and a 'dead' image. It reflects on the materiality and immateriality of images; particularly how digital reproduction influences our perception of history and culture. It also considers the preservation and understanding of cinematic spaces, promoting them as critical tools for constructing memory and identity. A new lens for examining the materiality of cinematic space is proposed through the integration of image theory, spatial theory, and film theory.

Prologue: Crisis and Reimagining Spaces

The era of affordable oil is ending. Deforestation and other industrial activities, marked by the wasteful use of resources and energy, coupled with the modernization of housing, have collectively led us to where we find ourselves today: in a state of emergency. Even though modern comfort was dependent on the availability of fossil fuels (Rahm, 2023, 11-16), today it is evident that we must diminish the amount of carbon dioxide emissions. The world's infrastructure was built for a climate that no longer exists. Acknowledging this is not sufficient; the question remains: How do we proceed? In *Manifestes 6: The Anthropocene Style*, Philippe Rahm (2023) suggests an action of redefining our formal and material choices, and ultimately, our aesthetic, cultural and social values. An environmental emergency necessitates a reconsideration of the use of space and materials. In cinema, the portrayal of crises often leads to the breakdown of conventional spatial representations. 'Liquid cinematic spaces' frequently emerge in response to crisis or transformation, whether environmental, social or psychological, creating fluid, hybrid spaces that challenge traditional forms.

Cinematic Space as Fluid and Dynamic: Architecture of Fragments and Figments

> *Fluids travel easily. They 'flow', 'spill', 'run out', 'splash', 'pour over', 'leak', 'flood', 'spray', 'drip', 'seep', 'ooze'; unlike solid, they are not easily stopped – they pass around some obstacles dissolve some others and bore or soak their way through others still* (Bauman, 2000, 2).

This text explores the concept of 'liquid cinematic spaces', examining how modern cinema has evolved to present fluid, flexible spaces that disrupt conventional notions of spatial representation. The concept of fluid cinematic space encompasses not only the presentation of physical environments, but also the liberation of narrative structures and temporal sequences from traditional linear or static frameworks. It initiates a discussion on interpreting cinematic space in light of climate change and the digital era, proposing it as a counter-archive – an alternative form of historical and cultural documentation distinct from traditional archives. Examining its potential to become a valid digital asset requires a multifaceted approach. The key question is: how is cinematic space constructed, perceived and potentially reinterpreted?

By examining how modern cinema engages with and redefines space, such spaces can resist, challenge and even rewrite dominant narratives. Each section of this essay looks at a specific facet of cinematic space, from its visual representations to its role in reshaping memory and history.

Introducing the concept of 'liquid modernity,' Zygmunt Bauman (2000) argues that the contemporary era represents a continuation or late development of early modernity. Instead of 'postmodern', Bauman proposes the term 'liquid' to capture the state of continuous flux and displacement within contemporary societies. Accordingly, modernity has inherently been dichotomous. In the first case, it signifies a concept of perfection, order, prediction and manageability – 'the idea of a final destination'. Yet it embraces change as continual transformation, with the aim of dismantling old concepts and traditions. The nature of change itself is the primary distinction between the initial 'solid' and the contemporary 'liquid' phase of modernity.

Cinematic space, whether imagined, real or hybrid, exists as a nexus between constructed and lived realities, intertwining elements of the mundane with the extraordinary. We can perceive it as an architecture of fragments and figments – things believed to be real but existing only in imagination – especially given that cinematic spaces are primarily constructed through formal or physical montage techniques. Film director and scholar Éric Rohmer identifies three types of space in film: *image space* (the spatial conditions and characteristics within a single shot), *architectural space* (the arrangement of landscape, buildings, and objects prior to filming), and *cinematic space* (the virtual space constructed by the viewer's imagination using the fragmentary elements provided by the film) (Kretzer, 2021, 43). It encapsulates how cinema creates a unified, immersive world by synthesizing these elements into a cohesive whole. The term cinematic space also refers to the imaginary environment that viewers construct through the film's narrative and montage. Additionally, this space forms in the viewer's mind as they connect various shots and scenes, relying on their interpretation and perception of space. The film's editing, continuity and spatial relationships shape this perception.

'Liquid cinematic spaces' represent constant change and instability. Similarly, the depletion of fossil fuels reflects a shift in global stability, where our resources, energy and spaces are no longer secure, stable or 'solid'. This mirrors how contemporary films portray spaces that flow, evolve or transform. The fluidity of space in cinema challenges human dominance over space. Liquid cinematic spaces also reflect a world where human control is no longer absolute, and environments become dynamic, responding to forces beyond human intervention. This relates to the suggestion that we must rethink our values in the face of environmental collapse.

To deeply analyze cinematic space, one should understand the image and medium of film itself, its historical development and the technological impacts that have shaped our perception. A starting point entails synthesizing different theories of image, space and film and applying them to specific examples. This approach, by rethinking the practice of archiving cinematic architecture, interiors and landscapes and expanding it into the digital realms, aims to refresh our perception of something familiar. This, in turn, could potentially lead to entirely new directions of inquiry.

Space versus Place

> *While places are constantly creating and molding our identities, spaces allow us to examine the identities which have been made* (Feireiss, 2019, 17).

A fundamental aspect of humanity's existential nature, space serves as a crucial prerequisite for our exploration of the world. 'It serves as the storage media of our historical situatedness and testimonial of the cultural fabric in which we live' (*ibid.*). In everyday language, we frequently use the terms 'space' and 'place' interchangeably. Both represent distinct aspects of our spatial experiences and interact dynamically across various fields of study. Lukas Feireiss (2019) defines a space as an abstract concept, a blank slate that awaits personal experiences to imbue it with meaning. A place is 'a location created by human experiences', and can be seen as a space that has acquired meaning through human perception and interaction.

Space is more abstract and general, whereas place is specific and filled with cultural, social and personal meanings. Understanding the differences between space and place enriches our comprehension of how individuals and societies perceive, experience and utilize their environments.

Fixed and Stable, Peripatetic and Fragile Spaces

> *Everything participates in the reality of spaces. Animals and plants, natural sceneries, colors, fragrances, taste, language, sex, and money have spatial effects, too* (Baier, 2020, 86).

Franz Xaver Baier developed a significant and comprehensive theory of 'lived space', distinguishing various types of experiences within spatial contexts. This radical theory of space highlights the subjective and experiential aspects of architecture and urban environments. It provides a framework for understanding how individuals experience and interact with their spatial surroundings, integrating sensory, emotional and cognitive dimensions. Baier (2020, 84-98) presents new phenomenological viewpoints on architecture, art and various other domains of experience, setting them apart:

1. *Existential space* is inseparable from objects and places; it emerges as a result of existential interrelations.
2. *Space in leaps and bounds: fractal geometry of lived space*; here, 'approximations and distances' are more significant than 'strict positions and definitions.'
3. *Situational space: the most sensitive space*, which mediates universal reality with our specific circumstances.
4. *Meaningful space: space of total presence* encourages a perspective that values openness, exploration, adaptability and a deep awareness of the vast and complex web of relationships that shape our existence.
5. *Perceived space* highlights the active, subjective and dynamic nature of perceiving and constructing reality.
6. *Self-similar space* invites contemplation on how we perceive and navigate spatial relationships across different domains of existence, from the physical to the abstract.
7. *Invisible space: interior reality* embodies the concept that real-life space can be invisible, concealed, camouflaged or protected by visible reality.
8. *Virulent space: embracing chaos* advocates for a shift in perception, from seeking rigid order to embracing the chaotic and dynamic aspects of life.
9. *Peripatetic space: individual small space* is an intimate space that possesses both interiority and exteriority and can be activated by humming, crooning or whistling.

Our lives consist of numerous life spaces coexisting simultaneously. These spaces vary, sometimes remaining fixed and stable, while at other times they are peripatetic and fragile. All phenomena are defined by spatial aspects and simultaneously result in spatial consequences (Baier, 2020). Cinematic spaces, which reflect and reinterpret 'lived spaces', engage audiences emotionally, intellectually and aesthetically. They enable viewers to experience and contemplate various aspects of human existence and the environments we inhabit, whether real, imagined or somewhere in between.

Space is Cinematic: The Complexity of Cinematic Space

When we talk about space in films, we are essentially referring to 'cinematic space'. The assumption is that cinematic space, in all its complexity, heavily relies on 'scalelessness and relational thinking', which are key aspects of contemporaneity (Fernández Contreras, 2021). In today's context, where engagement with the constructed or built environment, public or private space, landscape or interior, room or house, increasingly relies on mediation rather than direct experience, montage has the potential to enhance or even facilitate a 'performative aspect of space'.

By blending artifice and realism, cinematic space reveals the present as a synthesis of past experiences. Cinematic space conveys stories through spatial manipulation, often contrary to the usual understanding of built architecture (Martin, 2014, 8-9) – diverging from traditional architectural interpretations and challenging conventional perceptions of architectural design. Furthermore, not only editing *techniques* shape meanings in cinema, but also *mise-en-scène* and cinematography, each contributing to the formation of cultural narratives.

The spatial dynamics created by the montage in the film closely exemplify Bauman's concept of 'liquid modernity' in space. One could argue that Russian avant-garde filmmakers such as Dziga Vertov (*Man with the Movie Camera*, 1929) and Sergei Eisenstein (*Battleship Potemkin*, 1925) were already intuitively engaging with this notion through their experimental approaches to montage. Their approach deeply investigates the interplay of space, time and representation. As a result, we can identify cinematic space as 'liquid space', enabling us to explore the various forms and dispositions it adopts.

According to Jacques Lévy (2013), cinema encompasses a variety of spatial languages, primarily composed of images. Lévy asserts that cinematic space manifests the coexistence of 'space as an environment' and 'spatiality as action' – geographicity. Lévy discusses the four types of cinematic spaces, primarily identified by André Gardies (1933, 9-14): 1. 'Cinematographic space' refers to the 'institutional' setting, such as a movie theatre or domestic environments with screens, that immerses or exposes viewers; 2. 'Diegetic space' refers to the reality that the film constructs, independent of the story – the film's geographicity, a space that serves as a set or context – also known as 'anecdotal space'. 3. 'Narrative space' relates to the specific spatiality of the characters, which gives substance to the story, as a framework for the action. 4. 'Viewer space' is a subjective space, the spatiality produced by the mode of communication that the film adopts towards the viewer. While all these diverse types of space shape our cinematic experience and serve as a compelling domain for experimentation, diegetic space – space as an environment – stands out for its potential for further exploration (Figure 1).

The relationship between space and cinema is inherently ambiguous. Despite space being omnipresent in film, it often goes unnoticed or remains invisible, unless the viewer becomes fully immersed in the narrative.

Cinematic space as heritage treats film as a living medium that challenges our experience of space. By depicting spaces through fictional narratives, film adds layers of meaning to public and private locations, linking them to cultural memory and collective identity. This reframing influences how we celebrate and preserve public and private spaces, particularly in a post-digital age where fiction often replaces truth, blurring reality and representation. In this context, moving images transform everyday mundane places into sites of memory, making cinematic space an active counter-archive. Cinematic spaces thus encourage us to reflect on the memories we might pass on to future generations and how these curated spaces form or contribute to our cultural legacy.

The omnipresence of moving images also reshapes our engagement with our surroundings. In a post-digital landscape dominated by screens – on phones, walls and urban installations – these projections create spaces that are both real and imagined. Such representations influence our perception of space, inviting us to inhabit environments enriched by personal memories, collective narratives and digital traces.

The relationship between cinematic space and architecture is multifaceted and dynamic. Films act as a medium through which spatial experiences are performed and reimagined. The fluid, liquid nature of cinematic space – combined with montage and the interplay of different spatial languages, as articulated by Lévy (2013) – provides a platform to challenge traditional architectural narratives and explore new possibilities in spatial representation.

Memory, Space and the Fluidity of History

The recognition of filmic architectural heritage is less evident than that of film as a form of intangible cultural heritage. If *historia magistra vitae* and film preserves the entirety of our outer reality – our 'collective subjectivity' – then we could use cinematic spaces to (re)construct memory for the digital archives of our future (Bačun, 2023). Given that the archive's fundamental characteristics include fragmentation and unpredictability, it is straightforward to draw parallels with cinema. However, wasn't the traditional archive originally created with fragments and figments of reality?

Normally, we perceive an archive as something passive – a field of dormancy, a place of storage for something at rest. Traditional archives, whether physical or digital, serve as repositories of history but are often constrained by institutional bias. In contrast, the fluid, subjective nature of cinematic space offers a stark departure from the more objective, static nature of traditional archives. Cinema's inherent flexibility allows for a more nuanced exploration of historical and cultural issues. Cinematic spaces provide an alternative, enabling filmmakers to rewrite, resist or reimagine historical narratives. Films that present alternative histories or reframe historical events through subjective experiences or fictionalized spaces allow cinema to function as a counter-archive. Cinematic recreations of historical events act as counter-memory, challenging dominant narratives through imaginative reconstructions of space and time.

These spaces complicate the notion of what constitutes an archive, suggesting that memory and history are not fixed but fluid, much like the spaces that cinema creates. Cinematic space is empowered to destabilize our understanding of history. Films that blend past, present and future – such as *Hiroshima, Mon Amour* by Alain Resnais (1959) – disrupt conventional temporal boundaries, challenging the idea that history follows a linear progression (Figure 2). This approach not only depicts the unfolding of events but also shapes the construction, reconstruction and deconstruction of memory and history. It emphasizes that the interpretation of history is subjective, shaped by the spaces we inhabit, both physically and cinematically.

The spatiality of an archive is another compelling topic. What if cinematic space could embody that specificity and become a metaphor for a city, a library or something more? Resnais' (1956) seminal essay film *Toute la mémoire du monde* (*All the Memory in the World*) exemplifies the relationship between an archive and its content (Figure 3). Resnais constructs a unique cinematic space by portraying the library as both protagonist and metaphor for collective memory, thus facilitating the process of remembering. He emphasizes the power of representation in the affective memory of images. This context views remembering as a performative act.

We can view history from a variety of angles, offering different perspectives on the phenomenon of memory and how we construct it. Indeed, what happens when one goes beyond the confines of an official archive, engages previously overlooked, unrecognized and unspoken sources, and broadens the focus from the individual to the collective?

An intriguing perspective on archiving comes from professor of film studies Paula Amad (2010, 96-133), who interprets the emergence of cinematography as an example of a home cinema with no utilitarian purpose. It 'merely exists to be forgotten, due to its focus on the incidental, the everyday, the non-essential information' (Horak, 2011, para. 3). This means questioning the coherence and objectivity of traditional archives and challenging the value of an archived item – whether its significance is historical or merely derived from its long-term preservation. The question is: can collaborative practices and subjectivities of cinema serve as a counter-narrative to an archive?

The collective voice could act as a curator, as it already does on resilient collective outsourcing systems like The Pirate Bay – a BitTorrent site – and various social media platforms such as Facebook, Instagram, TikTok, YouTube, etc. These platforms, acting as 'private' archives, can become influential in determining what is valuable to preserve and thus mark the end of institutional cinematography as we knew it (Bačun, 2023). This deterritorialization arrangement in relationships could potentially address the question: who holds authority?

A 'Living' or a 'Dead' Image: The Constantly Evolving Image of a Film

> *What is seen on the film screen, in the present moment of projection, is simultaneously the past; thus, the world's film material is transformed into a cinematic event that is always already in the past* (Šakić, 2016, 16).

While film is often regarded as a 'living image' capturing and reflecting the past and history, it can also be seen as a 'dead image' from the moment of its creation. Film theorist Dušan Stojanović (2009, as cited in Šakić, 2016, 17), influenced by André Bazin, describes film as a composition of moving images of deceased individuals; it revives something that has died – something that was once reality and now reappears before our eyes. In this sense, cinematic space consists of dead buildings that either existed in reality once or never did.

In the book *Cool Memories II* (1987-1990), Jean Baudrillard discusses the freeze-frame in cinema, which captures and arrests the entire movement of the city. Unlike photography, which freezes and preserves a moment of time, 'the cinematic image . . . through the addition of movement, is considered to represent the very unfolding of time, thus giving the illusion of the same duration as our experience' (Orlow, 2007, 178-179).

Baudrillard's distinction between film and photography highlights how the temporal qualities of each medium shape our experience of time, memory and the representation of reality. The film's 'ever-changing image', portraying the vitality of the present, contrasts with the photograph's static representation of the past and mortality, reflecting deeper philosophical insights into the nature of images in contemporary society. This concept suggests that films constantly transform and evolve, reflecting the fluid nature of reality and perception. We can apply it to analyze how films function as cultural artefacts, constantly reshaping our understanding of reality and memory through their evolving visual and narrative forms.

Film as a Medium for Constructing Memory with the Image's Immateriality

> *When one knows that something will soon be removed from one's gaze, that thing becomes an image* (Benjamin, 2006, 115).

The philosopher Henri Bergson, diverging from a purely materialistic view of the world, introduces the concept of reality as comprising not only physical objects but also images (Bergson, 1911). These images serve as mediators of collective memory, referring to the shared experiences, traditions and cultural narratives that shape a community or society's identity and understanding of the world. Bergson's non-traditional perspective enriches our understanding of how subjective experience and memory contribute to our perception of reality and our place within it. In other words, Bergson speaks of the interconnectedness of human experience beyond empirical observation.

Reproducible technologies like film and photography have profoundly altered the relationship between history and materiality. Before the advent of digitalization, Walter Benjamin (2008) argued that the reproduction of images threatened the authority of the 'object'. However, in the digital era, the exponential growth in the production and reproduction of virally shared and endlessly reproduced imagery has changed. In contrast to Benjamin, artist and critic Hito Steyerl associates the value of the reproducible image with its 'usability'. With each download, post, like and share, the value of the image grows, establishing a new approach to dissemination and archiving. In digital environments, Steyerl (2010; 2012, 52) addresses the concept of decay symptoms:

> It is a complete mystification to think of the digital image as a shiny immortal clone of itself. On the contrary, not even the digital image is outside history. It bears the bruises of its crashes with politics and violence. ... The bruises of images are its glitches and artifacts, the traces of its rips and transfers.

Steyerl explores the concept of objectivity and examines how images alter the relationship between subject and object in identification. She shifts perspective. In her essayistic video, *How Not to Be Seen: A Fucking Didactic Educational.MOV File* (Steyerl, 2013), inspired by a Monty Python sketch from 1970, she teaches various strategies for remaining 'unseen' in the digital environment. Steyerl's concepts offer a new approach to digital archives and digital heritage, providing a fresh perspective on materiality in digital environments.

Additionally, theorist, writer and curator Claudia Giannetti (2021, 66-93) introduces the concept of the 'ecology of images and media' to illustrate how our visual environment is saturated with 'hyper-representations' – highly mediated, filtered and manipulated images that claim to reflect reality but instead distort it. These 'hyper-representations', constructed through layers of filtering, editing and curation, present a version of reality that acts as a fragmented mirror. In our daily interactions, we often fail to recognize how they shape and alter our understanding of the world. Giannetti's concept challenges us to rethink the 'truths' conveyed in our image-saturated world, urging us to question the reliability and interpretations of media content sources. The 'ecology of images and media' invites us to examine how multimedia affects our understanding of contemporary issues, events and even our identities. It encourages us to critically reflect on the media shaping our worldview, distinguishing between generated narratives and the complex realities they represent.

Film Experience: Cinematic Space as a Sequence of Images

Chantal Akerman reflects on the perception of time in cinema:

> *When you read a text, you're in your own time. That is not the case in film. In fact, in film, you're dominated by my time. But time is different for everyone. Five minutes isn't the same thing for you as it is for me...* (Rosen, 2004).

Ronald Barthes notices: 'Do I add, to the images in movies? I don't think so; I don't have time: in front of the screen, I am not free to shut my eyes; otherwise, opening them again, I would not discover the same image' (Barthes, 1981, 55). Seeing a film half a century ago was a different experience than it is today, with the ability to manipulate, stop, fast-forward, rewind and replay movies easily at will. New technologies, particularly the advent of video recorders, DVDs, personal computers and smartphones, recontextualize all media in various ways. According to David Company (2007), the complete film experience, with its particular way of viewing known as 'the integral film', is over, despite cinema studies' efforts to preserve it. New viewing techniques are reshaping this 'old object'. Cinema is no longer exclusive to specialists; it has entered the domestic sphere.

Jacques Rancière (2009) asserts the emancipation of the spectator. Today, spectators, unlinked from the physical space of the cinema, engage with the medium as active participants and can manipulate it more easily than ever before. Time unfolds before and after the cinematic experience, especially through virtual extensions on social media, where they 'mix, remix and republish visual content in a constant flow of images' (Duarte, 2022, 11).

In the post-digital era, moving images are in our hands, on our wrists, in our pockets and on walls that have turned into screens – essentially, everywhere around us.

Victor Burgin (2004), drawing on Michel Foucault's lecture 'On Other Spaces' (Foucault, 1986), refers to 'cinematic heterotopia', a network of distinct yet overlapping interfaces and viewing habits where images undergo delays and replays, storage and advertising and occasionally serve as replacements for actual images from memory. According to Burgin, memory is spontaneously organized into a succession of short-sequence images through interactions between an individual and their cinematic environment, creating a narrative. He provides a compelling perspective on revisiting cinematic spaces methodically, seeing them as sequences of images, especially with the concept of the still image, which is closer to the process of remembering than film. Burgin's concept highlights how effectively still images capture the operation of memory, often storing moments as vivid, fragmented sensations rather than seamless, fluid sequences. We engage with cinematic spaces in a way that allows each frame to convey its own mood, emotion and spatial significance by seeing film as a collection of distinct images rather than a continuous narrative. Within this methodological lens, each image uniquely contributes to the spatial and emotional layers of the story, enabling a more thorough analysis of how film constructs meaning. This process deepens our understanding of how films affect and resonate within the viewer's memory landscape, allowing cinematic spaces to transcend their immediate on-screen representation and embed themselves in personal memory as a series of powerful, isolated experiences.

Our understanding of film shifts with this analytical approach, which reveals how each frame's spatial composition and emotional undertones combine to create a comprehensive, memory-like experience that is both intensely personal and universally accessible. By analyzing a film as a series of images, we can explore the multifaceted nature of spatial construction, narrative development and the emotional resonance intrinsic to cinematic storytelling. This method not only enhances our understanding of film as an art form but also enriches our appreciation for how cinematic spaces unfold and resonate with viewers and their environments.

Conclusion: Liquid Cinematic Space as a Counter-Archive

Our existence unfolds across both space and time, compelling us to consider ourselves within these interwoven realms that continuously shape our perception and understanding of the world. Instead of providing definitive answers, this exploration suggests that space is fluid, time is an asterisk, and daily life remains unpredictable – especially in the context of climate change. Recognizing how technology has historically framed society enables us to better anticipate how current cinematic developments might influence the future. In this context, the act of revealing – rather than merely conveying a message -adds critical value to both analysis and interpretation.

The essence of cinema lies not merely in its capacity to reproduce physical reality, but in its potential to interpret, reconstruct, and reimagine that reality in meaningful ways. In this regard, film operates not as a passive observer but as an active agent – inviting viewers to experience reality as a dynamic construct shaped by the interplay of image, narrative, and emotion.

When perceived as 'liquid,' cinematic space becomes a site where traditional structures of memory and history are dismantled and reimagined. Its ability to challenge conventional notions of space, time, and memory allows cinematic space to emerge as a counter-archive. These fluid spaces invite audiences to engage with history and memory in subjective, fragmented ways. By providing alternative visual and narrative representations of the past, cinema challenges the rigidity of traditional archives. In doing so, it empowers cinematic spaces to question and redefine dominant historical narratives, preserving overlooked or marginalized experiences and voices. This transformative potential enables cinema to turn the fixed into the fluid, reconfiguring spatial records into living, evolving forms of cultural memory. Through the study of these cinematic spaces, we may uncover new methods for archiving memory-methods that honour both heritage and the future – while deepening our understanding of how profoundly moving images shape our environments.

When analyzed through the lens of 'liquid modernity,' cinematic space reveals its mutable nature, mirroring contemporary uncertainties marked by climate change and accelerated technological development. As we navigate this mediated ecology, critical engagement becomes essential for understanding how such spaces influence identity, memory, and the narratives that link our past to the futures we imagine.

In contemporary cinema, spatial representation frequently mirrors the complexity and instability of modern life. Employing computer-generated technology (CGI), fragmented editing, and non-linear storytelling, filmmakers transcend static forms to emphasize movement, transformation, and unpredictability. These cinematic spaces open up new narrative possibilities, offering immersive and multifaceted experiences. By embracing fluid environments, filmmakers challenge traditional spatial boundaries, crafting worlds that blur the distinctions between reality and imagination, the physical and the virtual – both in the past and today.

Although film inherently captures life's motion and resists immobility, analyzing it through still images remains a productive and insightful method. Isolating individual frames allows us to pause the flow of time and examine them as discrete compositions. Each still encapsulates a moment within the broader narrative, revealing deliberate choices in framing, lighting, and symbolism that might otherwise be overlooked in the continuous movement of a sequence.

While some films present information too explicitly – leaving little space for discovery – others invite viewers to notice subtle details and embrace multiple interpretations. These films, in which images stimulate critical reflection rather than overwhelm, are particularly valuable for archival preservation, regardless of genre.

Though it may seem as if we have seen it all before, revisiting cinematic spaces with fresh eyes – setting aside prior knowledge and cultural biases – can reveal new insights. By accepting the fragility, transience, and vulnerability of ever-changing spaces, we become more attuned to their dynamic nature, thereby deepening our appreciation of them. This shift in perspective can strengthen our relationship with the environment: welcoming change and impermanence fosters a heightened sensitivity to the delicate balance that shapes both cinematic and real-world spaces. In doing so, we are invited to rethink and adapt our values, acknowledging and embracing the transient and interconnected nature of the world around us.

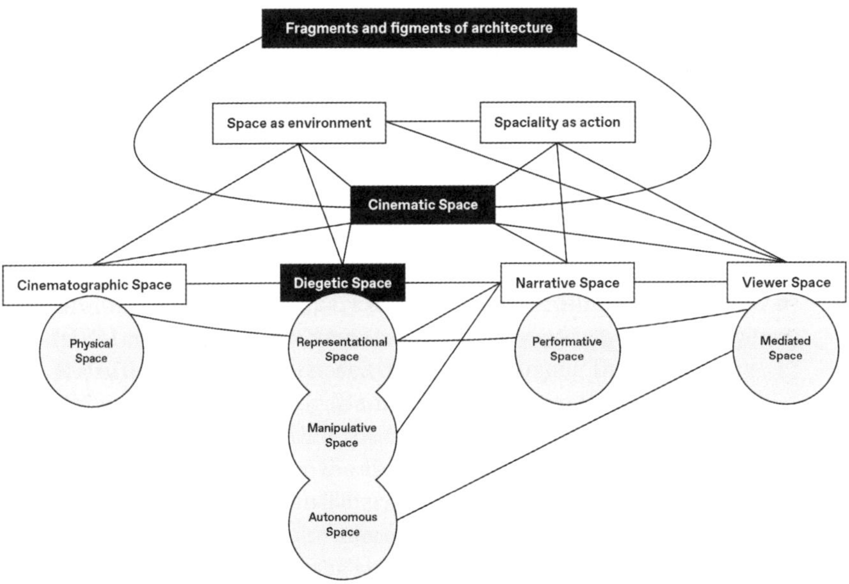

Figure 1: A graphic representation of André Gardies' four types of cinematic spaces, expanded through Jacques Lévy's concepts of space, with a particular focus on diegetic space, further enriched by Fernández Contreras' insights into the performative aspects of space. Nina Bačun, 2024.

Figure 2: Composition of still images from *Hiroshima, Mon Amour* (Resnais, 1959). © No further use allowed.

Figure 3: Composition of still images from *Toute la mémoire du monde* (Resnais, 1956). © No further use allowed.

References

Amad, P. (2010). *Counter-Archive: Film, the Everyday, and Albert Kahn's Archives de la Planète* (New York, NJ: Columbia University Press).

Bačun, N. (2023). 'Arhitektonski filmski prostori kao proto-arhiv kolektivne memorije', *Serbian Architectural Journal, 15*(3), 288-313. https://doi.org/10.5937/saj2303288B

Baier, F. X. (2020). 'Lived space', in L. Feireiss (Ed.), *Space is the place: Current reflections on art and architecture*, 84-98. (Leipzig: Spector Books / BNKR).

Baltazar, M. J., & Quadros, T. S. (eds.). (2021). *Image in the Post-Millennium: Mediation, Process and Critical Tension* (Portugal: Esad/idea and Onomatopee).

Barthes, R. (1981). *Camera Lucida: Reflections on Photography* (transl. R. Howard). (New York, NJ: Hill and Wang).

Bauman, Z. (2000). *Liquid Modernity* (Cambridge: Polity Press).

Benjamin, W. (2006). *The Writer of Modern Life: Essays on Baudelaire* (ed. M. W. Jennings; transl. H. Eiland, E. Jephcott, R. Livingston, & H. Zohn). (Cambridge, Massachusetts and London: The Belknap Press of Harvard University Press).

Benjamin, W. (2008) [1935]. *The Work of Art in the Age of Mechanical Reproduction* (London: Penguin Books).

Bergson, H. (1911) [1896]. *Matter and Memory* (transl. N. M. Paul & W. S. Palmer). (London: George Allen and Unwin).

Burgin, V. (2005). *The Remembered Film* (London: Reaktion Books).

Company, D. (2007). *Cinematic: Documents on Contemporary Art* (London and Cambridge, MA: Whitechapel Gallery and MIT Press).

Duarte, G., Cordin, G., & Ferrando, D. T. (eds.). (2022). *Expanding Spatial Narratives* (Milan: Mousse Publishing).

Feireiss, L. (2020). *Space is the Place: Current Reflections on Art and Architecture* (Leipzig: Spector Books).

Fernandez Contreras, J. (2021). *Manifestes 1. Manifesto of Interiors: Thinking in the Expanded Media* (Geneva: HEAD Publishing).

Foucault, M. (1986) [1967]. 'Of Other Spaces' (ed. J. Miskowiec), *Diacritics, 16*(1), 22-27. https://doi.org/10.2307/464648

Gardiès, A. (1993). *L'espace au cinema* (Paris: Méridiens Klinck-Sieck).

Giannetti, C. (2021). 'Post representation: Towards a theory of hyper-representation and underrepresentation. Ecologies of the image and the media', in M. J. Baltazar & T. S. Quadros (eds.), *Image in the post-millennium: Mediation, process and critical tension*, 66-93. (Eindhoven: ESAD/idea and Onomatopee).

Horak, J.-C. (2011). *Counter-archive: Film, the everyday, and Albert Kahn's Archives de la Planète* [Review of the book *Counter-archive: Film, the everyday, and Albert Kahn's Archives de la Planète*]. *Screening the Past*. Available: http://www.screeningthepast.com/issue-32-reviews/counter-archive-film-the-everyday-and-albert-kahn%E2%80%99s-archives-de-la-planete/ (accessed 9 April 2025).

Lévy, J. (2013). 'On Space in Cinema', *Annales de Géographie, 694*(6), 689-711. Available: https://www.cairn-int.info/article-E_AG_694_0689–on-space-in-cinema.htm (accessed 9 April 2025).

Kretzer, A. (2021). 'Beyond the Haze of Carnival Candles: Cinematic Space in Architectural Design Education', in E. Garip & S. B. Garip (eds.), *Handbook of research on methodologies for design and production practices in interior architecture*, 23-43. (IGI Global). https://doi.org/10.4018/978-1-7998-7254-2.ch002. Available: https://www.irma-international.org/chapter/beyond-the-haze-of-carnival-candles/265760/ (accessed 9 April 2025).

Martin, R. (2014). *The Architecture of David Lynch* (London: Bloomsbury Publishing).

Orlow, U. (2007). 'The dialectical image: *La Jetée* and photography as cinema', in D. Company (ed.), *Cinematic: Documents on contemporary art*, 178-179. (London & Cambridge, MA: Whitechapel Gallery and MIT Press).

Rahm, P. (2023). *Manifestes 6. The Anthropocene Style* (Geneva: HEAD Publishing).

Rancière, J. (2009). *The Emancipated Spectator* (transl. G. Elliott). (London: Verso Books).

Resnais, A. (director, 1965). *Toute la mémoire du monde* [film]. Les Films de la Pléiade.

Resnais, A. (director, 1959). *Hiroshima, mon amour* [film]. Argos Films/Como Films.

Rosen, M. (2004). *In her own time: An interview with Chantal Akerman. Artforum*, 42(8). Available: https://www.artforum.com/features/in-her-own-time-an-interview-with-chantal-akerman-168431/ (accessed 9 April 2025).

Steyerl, H. (2010). *A thing like you and me*: *e-flux journal*, 15, 1-7. Available: https://www.e-flux.com/journal/15/61298/a-thing-like-you-and-me/ (accessed 9 April 2025).

Steyerl, H. (2012). *The wretched of the screen* (Berlin: Sternberg Press).

Steyerl, H. (director, 2013). *How Not to Be Seen: A Fucking Didactic Educational .MOV File* [film]. e-flux. Available: https://www.e-flux.com/film/384606/how-not-to-be-seen-a-fucking-didactic-educational-mov-file/ (accessed 9 April 2025).

Šakić, T. (2016). *Modernizam u hrvatskom igranom filmu* (Zagreb: Disput d.o.o.).

Rhythms of Rural Life Behind the Dikes

Peter Veer

Documentary films offer valuable insights into the societal and ecological impacts of large-scale water management projects, yet their narratives often reflect dominant ideological perspectives. This essay explores how historical Dutch documentaries framed rural life and environmental transformations, focusing on three government-distributed films: *Eiland van vertrouwen* (*Island of Faith;* Fernhout, 1950), *Luctor et Emergo* (*I Struggle and Emerge;* Van Essen, 1956), and *Van oud naar nieuw* (*From the Old to the New;* Penning, 1957).

Drawing from theories in film and cultural studies, I analyze these films using close viewing and audiovisual rhetoric methods to reveal their underlying messages. I introduce the concept of 'engineering films' – a genre that highlights technical processes while neglecting the lived experiences of local communities. Inspired by Henri Lefebvre's concept of rhythmanalysis, I reflect on the rhythms of rural life and the ways audiovisual narratives construct perceptions of landscape and modernization.

This interdisciplinary approach underscores the importance of critically engaging with historical media to understand past infrastructural projects and their long-term social and ecological consequences, bridging the gap between technical disciplines and the humanities.

Introduction

Documentaries can offer valuable insights into the societal and ecological implications of large-scale water management projects in the past, present and future. Like other media, films are full of information, profound or hidden messages and meanings. The languages of images and audiovisuals are very well equipped to transmit these cultural and ideological convictions (Hall, 1997). But how can the cultural meanings about social and physical environments in the films be unveiled?

In my doctoral thesis *Framing the Dutch Landscape* (Veer, 2020a and 2020b) I studied the film practises of the Dutch Ministry of Agriculture in the period from 1945-1985. The corpus is the extensive collection of 533 informative films distributed by the Film Unit of the Ministry. In the study I also analyzed the production, distribution and promotional practices connected to the films. In this essay I will use 'societal and ecological implications' as key concepts to elaborate on the films. I will focus on three productions in the collection of the Ministry: *Eiland van vertrouwen* (*Island of Faith*; Fernhout, 1950), *Luctor et Emergo* (*I Struggle and Emerge*; Van Essen, 1956), and *Van oud naar nieuw* (*From the Old to the New*; Penning, 1957). Using the last film as an example, I will explain an experimental research tool to unveil the hidden meanings from the storyline and in the audiovisual layers of a film (Hesling & Peters, 1985; Hall, 1997; Banks, 2001; Vos, 2004; Van Leeuwen, 2005; Rose, 2012). The work of the French sociologist and philosopher Henri Lefebvre (1992) on rhythmanalysis is a major source of inspiration for this essay. However, I begin with a personal anecdote.

Anecdote: Some Rhythms of Rural Life

On a cold and wet day in May 2024, I installed myself behind the computer. That afternoon, I dived further into the information about the TU Delft project 'Reporting space, time and everyday life in the delta'. The Instagram account 'Reporting the Delta' (2024) gives an impression of about 65 films studied in the project. One reel in particular drew my intention: *Als het water stijgt* (*When the Water Rises*) from 1975. This shows part of the story of one of the many reinforcements of the dike near Dodewaard, in the Betuwe region along the river Waal, the majestic main stream of the Rhine in the Dutch delta. I felt I needed more information about the historic context of that film, as the Waal was personally important to me.

When I was young in the 1960s and 70s, I cycled over this dike very frequently (Figure 1). My parents were fruit farmers with tens of thousands of apple, pear and cherry trees. My primary school was in the village nearby and every morning and afternoon my siblings and I biked along the river, all year around. In early spring, when the snow in the Alps melts and there is a lot of rain in the Upper Rhine Basin, the floodplains on the river side of the dike change from lowland meadows into an inland sea. With strong winds and heavy rains, the waves crashed almost to the top of the dike. Although we pedalled extra fast, we often arrived at school late, a bit breathless and soaking wet.

Later in the season, crops are sown, and calves, piglets and chicks are born. On the weekends we saw tourists in cars and buses admiring the fruit trees in full bloom. In the summer, all kinds of grain, hay and other crops were harvested. The water level in the river decreased, the freight ships came by only partially loaded, and we swam in a small lake that was the result of a historic dike breach. In the autumn, apples and pears were picked; the sugar beet harvest trucks with clay on their wheels made the roads slippery. Trees lost their leaves; the cattle were put in the stables. In cold winters we could skate on De Strang, a quiet ancient river branch just along the dike.

A few days later in 2024, I visited the village of my youth and cycled on the dike again. At first sight the rhythms of daily life had hardly changed over the previous decades. I took a picture on the same location where *Als het water stijgt* (Boelens, 1975) was shot about fifty years ago. I asked myself what I could contribute to the group of professionals who are managing the Instagram account. What else is needed, except a proper translation of the voiceover text, to understand the Dutch-language film? These questions are important to learn more from the films making up the rich historic corpus of Dutch documentaries. How can hidden significance and meanings be distilled out of these films, most of which are embedded in Dutch culture and regional history? What basic film analysis techniques could help scientists to better understand the films about the large-scale, technology-based infrastructural works and their impact on citizens and their daily lives in the past, present and the future?

The film *Als het water stijgt* is about the reinforcement of the dike in the first years of the 1970s. On Instagram only a short part of the film is presented. A male voiceover speaks in Dutch. The images change slowly. They are all wide-angle; no close-ups are shown. The focus of the images is on large excavators and diggers, on piles of sand and clay; no people are portrayed. The represented space is a construction site and the storyline shows the rhythms of the large machinery and their operators.

Due to my personal connection to this place, I can tell lots of stories about the cultural and historical context of *Als het water stijgt*, such as the local impact of dike breaches in former centuries. I know the stories I was told about the floods in the low-lying centre of the Betuwe during the winter of 1944, when the region was a battlefield after the failure of the Market Garden operations. I know what the dike looked like when I was a boy.

In *Als het water stijgt*, the experienced space and time of the local community in that period is hardly represented. The fragments highlight efforts to prevent future dike breaches and detail the technical execution of the reinforcement works. No images are shown of people living in the region, how they used the dike, no stories told about what houses were demolished; no inkling that the works caused traffic congestion for many months because the dike was the only connection to other villages and regional towns. I know there were also positive consequences: the Betuwe became a safer place due to reducing the risk of flooding, the reallocation of the local garbage dump, and the new, wider dike could accommodate the growing traffic flow better. So, the film gives hardly any clues about the impact of the large-scale water management projects on the dike on the local people. Due to the reinforcement works the old river branch De Strang, a unique ecosystem, was redesigned. In the 1970s the impact of public works on the natural environment was a minor issue. That is probably why *Als het water stijgt* gives no insight in the consequences of the works for the local ecology.

To learn more about the societal and ecological implications of the dike reinforcement I could suggest other films, which show why and how dike reinforcements were conducted, but also portray the social resistance that such projects evoked in local communities. A selection can be found on the *Mijn Gelderland* website (2024). Other films give insights into the events in 1995, when the water rose to extremely high levels, the river dikes almost broke and 250,000 people in the region had to evacuate from their homes and farms (NOS, 2020). Other dike reinforcements are now planned by the regional water board Waterschap Rivierenland. Yet, *Als het water stijgt* gives me a starting point to elaborate on what can be learned from historical documentary films about everyday rural life in the Dutch delta.

Films in Their Cultural-Historical Context

My first viewing of *Als het water stijgt* (Boelens, 1975) as a film historian raised many questions. Who were the commissioner and the filmmaker and what audience did they have in mind? What was the message they intended to convey to their public? What did this movie contribute to the specific societal debates about dike reinforcements in the 1970s? To answer these questions, we may dive into the discipline of film studies.

Basically, the medium of film consists of two simultaneously presented lines of information: moving images and sounds. Both lines compose a third layer, that of meaning. Filmmakers combine the creative arts, crafts and techniques of image and sound design and of cinematographic storytelling. Film is a technology-based, modern medium, invented almost at the same time at the end of the nineteenth century in France, the United Kingdom and the United States. From the beginning of the film, it was obvious that it would have significant impacts on societies, norms and values of people all over the world. According to a famous anecdote, at the first presentations of the film *L'arrivée d'un train en gare de La Ciotat* (Lumière & Lumière 1895) in Paris, the audience, out of fear the train would run over them, screamed out and hid themselves on the floor.

Genres in film history are groups of audiovisuals with for instance common production techniques, corresponding forms, styles, similar themes, subjects or likewise purposes. In film there are two main genres with no strict distinctive characterizations: 'fiction films' are mostly seen as a product of human fantasy and meant for entertaining and making profit. The starting point of 'non-fiction' or a documentary is in the real world; the experienced space and time can be far away or close by, in the past, present and future. These films are mostly used for information, education and communication. Still, the taxonomy in audiovisual genres is not well defined and the boundaries are vague. In film studies a documentary is mostly seen as an artistic piece of work that also reflects the creative vision of the author or maker (Monaco, 2009; Nichols, 2010).

The term 'documentary' was coined by the Scottish filmmaker, producer and researcher J. Grierson in 1926. An early application of the medium of (documentary) film in science was in anthropology. One of the few films he directed himself was *Drifters* (1929) about the work and living conditions of British herring fishermen. The rhythms in the montage of this film can be seen as a cinematographic translation of the difficult life of these people, on their boats, in the harbour, in the fish processing industry and during the distribution to the consumers in towns and cities. Still, documentary films could raise debates. The documentary *Nanook of the North* by Robert J. Flaherty (1922) is a famous but controversial example. It pretended to report on the everyday life of indigenous people in Canada. However, in contradiction to the anthropological paradigm and the claim of the film itself, some sequences were staged. In the montage, the filmmaker did not respect the order and rhythm in life of the Inuit people.

Film history shows that audiovisual stories can contribute to major social changes. An example is the use of films by the young Soviet Union in the 1920s and 30s to promote the 'New Man'. For instance, in *Old and New or The General Line* directed by Sergei Eisenstein (1929) about communist rural reform in Ukraine, Eisenstein developed a new film language based on montage and rhythms. Another example is the way in which Hollywood films still export the 'American Dream' across the world. Audio-visual media are thus the most dominant in spreading norms and values worldwide. Although many films are shown around the globe, there is no well-defined universal film language. Even when the film-makers and the audience live in the same time period and share the same dominant culture, there is room for debate and misunderstandings. The message the makers intend to communicate can differ from the meanings the audience creates in their minds.

An interesting historical perspective for this essay is that of 'user films'. This is a collective term for educational and instructional films, corporate and advertising productions. These genres have in common that the message is intended to be instrumental to the commissioner. Already in the 1920s, large companies in the United States such as the Ford Motor Corporation and the financial company Western Union started to use films to regulate work processes according to the ideas of business experts like Frederick W. Taylor (1856-1915). By showing films, the management wanted to familiarize the staff with the routines, norms and values that applied to the work in their companies. The aim was mainly to prevent mistakes. These films extended to marketing and information and also to non-commercial environments such as education, public information and healthcare. They are, in line with Foucault's philosophy, an institutionalized tool of cultural power to increase productivity in society (Hall, 1997, 44-51) In this way, audiovisuals are a very powerful medium to inform people, influence their attitude and even change norms and values in societies. Just like other cultural products, film is part of a societal exchange of ideas. A filmmaker, who often works on behalf of a commissioner, has the intention to contribute to a debate. In archive films, the production represents in one way or another the issues that were important in a specific region in a certain period of time (Hediger & Vondereau, 2009; Acland & Wasson, 2011; Hediger, 2024).

Als het water stijgt (Boelens, 1975) was kept in archives for almost fifty years. I presume that nowadays, I am one of the few people watching this production, as it recalls personal memories. Yet, the film will have a different meaning for the machinist shown in the film, the director of the regional water board or even my classmates at my primary school. And I am sure these meanings differ from the message the filmmaker intended to transmit. So, the combination of moving images and sounds makes films multi-interpretable. How to find clues about the significance of an image, a clip, a sequence, a film?

In cultural studies, close reading is a well-known technique to analyze written texts and draw conclusions about the meanings of the story that remain hidden in everyday reading and sketchy scanning of phrases. A similar 'close viewing' technique can be used when analyzing films. Close listening to the voiceover of *Als het water stijgt* makes clear that it does not use the regional dialect. It uses a grammatically passive way of speaking, is full of technical jargon and has an abstract tone. The fragment provides no information about the agent who was responsible for the reinforcement of the dike or who commissioned the film. The image line focuses on large machinery. The film does not include a social point of view and shows no empathy with the local inhabitants. The aim of the production is probably not to inform people in other villages about upcoming reinforcement projects. The film does not contribute to the serious societal debates going on in that period about the technocratic decision-making processes and working methods of dike reinforcement projects, which destroyed cultural heritage 50 kilometres downstream in the village of Brakel.

The voiceover and the image line give the impression that the filmmaker addresses their argument to technical engineers or other experts. My preliminary conclusion is that this film was commissioned by either the contractor of the excavation works or the technical unit of the regional water board responsible for the reinforcement of the dike. The intended audience would be colleagues, in order to convince them of the technical skills involved. The contractor was probably looking for new assignments.

Als het water stijgt can be seen as an example of an 'engineering film'. I will discuss this genre of audiovisual productions further in this essay. This is a genre of films that focus on the technical practices of specific working processes. The films in this genre were mostly commissioned by engineers, who saw colleagues in the sector as the target group. The aim of an engineering film is to display technical craftmanship. This genre is also present in the collection of informative films of the Dutch Ministry of Agriculture.

Dominant Land Use in the Dutch Delta: Agriculture

By nature, due to the fertile arable soils and the mild climate, land use in the Dutch delta is dominated by agriculture. To understand the rural life on the Dutch islands and in its polders, a discussion of farming and agricultural land is required. In the Netherlands, the Industrial Revolution started late around 1850. Steam engines made ships faster; trains opened up the countryside. The import of cheap grain from Ukraine and the Americas lifted the regional restrictions of the 'Malthusian trap'.

In the First World War the Netherlands was neutral; compared to Belgium, France or Germany the negative consequences of the war were relatively small. But after World War II, the country was ruined, and many people had been murdered or wounded. Society was in complete chaos. The urban population had suffered from severe hunger in the final winter of the war. Rural sociographers observed that in large parts of the countryside many peasants were very poor and lived in unhealthy conditions in huts and hovels. In 1945, the production of these small farms did not rise above subsistence; there was hardly any surplus to bring to the market. The agriculture production techniques were backward and could not guarantee enough food at low prices for an industrializing society (Bieleman, 2008; Blom, 2017; De Haas, 2013; Davids & 't Hart, 2023).

Although in the last eighty years the Netherlands urbanized and industrialized rapidly, in 2021 agriculture still used about 54% of the national surface. This is more than in most other countries in western Europe except Denmark. It is astonishing that this relatively small country is number two on the global list of agricultural exporters after the United States (World Bank, 2024). These remarkable facts are the results of intensive structural policies of the Dutch Ministry of Agriculture in the period from 1945-1985, roughly in the Delta Works era.

The first Dutch Minister of Agriculture after World War II was the social democrat, former farmer and resistance fighter Sicco Mansholt (1908-1995). From 1957 to 1973, he was the first European commissioner for agriculture. In the reconstruction period just after World War II, Mansholt introduced a series of integrated rural development policies. The original aims were to increase agricultural production to ensure enough food for Europeans after the famine in the winter of 1944-45, to ensure sufficient foreign exchange to repair the national war damage and to diminish poverty in the countryside. His major sources of inspiration were the New Deal policy of US President F. D. Roosevelt to combat the 1929 Great Depression and the Tennessee Valley rural development project (Ellwood, 1989). The Marshall Plan imported these ideas and promoted the 'American Rural Dream' in the Netherlands. Mansholt was the main promoter of the transition 'from traditional to modern'.

Land consolidation (*ruilverkaveling*) was an essential part of the government programme to implement these agricultural policies. Land consolidation consisted of readjustments and rearrangements of fragmented land parcels and their ownership in regional projects. These projects were also used to improve rural infrastructure, water management systems and physical living conditions. This 'rural makeover' made the increase of agricultural production possible. Land consolidation had significant social and ecological implications on approximately two-thirds of the rural area in the Netherlands. To put the policy in practice, the new discipline of rural civil engineering (*Cultuurtechniek*) was created at the Agricultural University in Wageningen. The researchers at Wageningen University developed the programs on behalf of the Ministry of Agriculture. Besides rural civil engineering, other disciplines like agricultural business administration, rural sociology, communication sciences and landscape architecture emerged and flourished. Specialists educated in Wageningen became the technical and social engineers charged with carrying out the land consolidation projects. The Land Consolidation Service (*Cultuurtechnische Dienst*), part of the Ministry of Agriculture, coordinated the national program and the regional projects. Local and regional authorities such as municipalities, provincial governments and water boards were involved. Landowners, farmers and later also tenants had voting rights regarding the projects, since, in order to carry out the regional projects, the government needed the cooperation of the agricultural communities (Andela, 2000; Bieleman, 2004).

During land consolidation projects all over the country, intensive communications took place to promote modern working practices and lifestyles for the farmers, their wives and families (Figure 2). New communication techniques developed in the USA, such as film, were implemented (Van den Ban, 1982; Rogers, 2003; Karel, 2005). At the Ministry of Agriculture, a special department, the film unit, was active from 1945 to 1985. In these four decades, the unit managed and distributed a total of 533 films. These films were shown all over the nation at meetings of farmers and their families to disseminate all kinds of ideas about efficient agricultural techniques and modern rural life. In probably all the villages along the dikes, on the islands and in the polders of the delta, the local people saw at least some of these productions. All these films tell us something about the changes in the rhythms in daily rural life in all the regions along the rivers and shores.

Due to the national land consolidation program, about two-thirds of the agricultural land underwent a complete makeover. The look and feel of large parts of the Dutch landscape of today are determined by the design principles of the programme to stimulate efficient agriculture production. Even in regions which most people nowadays appreciate as 'traditional' and picturesque, the layout of fields, roads, waterways and tree lines is relatively new. Good examples are the former islands in South-Holland and Zeeland and the land in the polders along the rivers Rhine, Waal and Meuse (Andela, 2000; Blom 2017).

Two Dutch Agricultural Films about the Disastrous 1953 Flood

In *Bewogen landschap* (Veer, 2020) I analyze two films made by the Ministry of Agriculture that explore the social and ecological implications of the Delta Works. They represent different approaches to filmmaking. The first film, *Eiland van vertrouwen* (Fernhout, 1950), is set on the island of Walcheren. After the opening sequence, a picturesque impression is given of a farmer ploughing his field by horse power. Images of traditional landscapes and rural life are shown. In 1944, the last year of World War II, for strategic reasons the island was deliberately flooded by the Allied Forces to liberate the port of Antwerp from nazi Germany. Although this flooding was man-made and only ravaged one island, the aftermath can be seen as a prelude to the disastrous 1953 storm floods that killed many more people and animals, and destroyed vast numbers of communities and large surfaces of arable land.

Eiland van vertrouwen focuses on the government policy to reclaim land from the sea and to redesign and re-parcel the agricultural plots. The film reports the experiences of two local farming families suffering and recovering from the flood. To make their land fertile again, they are guided by rural extension officers. They also give the farmers a perspective on the future as free agricultural entrepreneurs. The state creates a new landscape for the farmers to work efficiently, so they can contribute to the food supply of the Dutch and European population (Figure 3).

In the resolution of the story, the Maljaars family stays on the rehabilitated island. The Wisse family leases a new farm in the newly reclaimed Noordoostpolder (Northeast Polder) and leaves Walcheren. The rational design of the Northeast Polder is presented as most suitable, even ideal, for agricultural land use.

The film gives insight into how the two farming families cope with the impact of the flood of saltwater on their land. Set in the direct aftermath of World War II, ecological themes were not at the forefront of society's agenda. But the film pays attention to cultural heritage, to the destructive force of the saltwater, the healing power of the rain, the first harvests, the planting of trees and a surviving mouse. The civil servants and agricultural engineers leading the redevelopment of the island were well aware of ecological and heritage values (Andela, 2000). I assume they discussed their motives with the filmmaker. Looking back, the 'Walcheren approach' also can be seen as an unintended 'final rehearsal' for how the Dutch government dealt with the disastrous flooding in the south-west in 1953. In the 'country created by the Dutch / where in all regions / the voice of the water / with its endless disasters / is feared and obeyed' (Marsman, 1936), the floods caused a national trauma. The nation was in mourning, but the flooding also generated a great social connectedness and energy. Large-scale national and even international aid programmes were set up. The practical Delta Works started even before they were passed into law by Parliament in 1958. The law not only initiated the building of the protecting dams, it also put an end to the isolated island culture. In most smaller harbour towns, fishing economies died. The law set the nation-wide standards for flood safety. These norms were the motive for reinforcement works in all coastal areas and inwards along the rivers, such as in Dodewaard.

The second film, *Luctor et Emergo* (Van Essen, 1956) tells the story of the island of Schouwen-Duivenland just after the 1953 flood (Figures 4 and 5). The title is also the motto of the Zeeland province: 'I struggle and emerge' – from the sea, obviously. The opening sequence shows that the film was commissioned by Grontmij NV, a rural development company. The focus in this film is not on the farmers and their work and lives, but on the heroic technical enterprise that Grontmij conducted. The voiceover relates how soil rehabilitation took place. The images show large excavators and diggers, vans at work in piles of sand and clay. In this production, people are portrayed: hard-working blue-collar workmen and a few white-collar technical engineers. In the frames there are no women, no children, no farmers or other people who lived on the island before or after the floods.

Like in the resolution of *Eiland van vertrouwen,* the image line in the last sequence shows a representation of modern agriculture, followed by a 'traditional' representation: after the large combine harvesting machine has harvested the field, the straw is pulled up on a cart by manpower and driven to the farm by a horse. The voiceover says: 'now the time is coming up when the farmers of Schouwen-Duivenland will work on the land and have rich harvests from soils that were almost destroyed forever by our hereditary enemy, the water'. At the end the bombastic voice refers to the contribution that Grontmij made to *Luctor et Emergo.*

This film gives no impression of how the local people of Schouwen-Duivenland experienced the flood and the large-scale civil and water management engineering works. On the other hand, it shows the enormous enterprise that had to be conducted to give agriculture and rural life a new start. The film represents the rhythm of life of the civil engineers, the workers, the large machinery. Although the Delta Works as a whole can be seen as a large-scale ecological intervention, in *Luctor et Emergo* (Van Essen, 1956) the theme of how to deal with nature is absent. Like *Als het water stijgt* (Boelens, 1975) this film fits the category of what I describe as an engineering film. With the film, Grontmij shows its skills to its potential clients and the general public. The firm aims to increase their business in the reconstruction of the islands and polders in the delta.

Analysis of the Film *Van oud naar nieuw* (1957)

My intention in this section is to illustrate how an initial film analysis can be conducted by urban scientists and water management engineers. This approach provides a research strategy and methodological framework for extracting the historical societal and ecological implications of large-scale environmental interventions from archival films. A more comprehensive description of this cinematographic method, specifically for analyzing documentaries, informational and promotional films, is provided in Veer (2020a and 2020b).

This rhetorical analysis method is designed to reveal meanings within an audiovisual work that remain unnoticed during casual viewings, whether in cinemas or educational and informational settings. As a case study, I chose *Van oud naar nieuw* (*From the Old to the New*; Penning, 1957), a key production from the Ministry of Agriculture's promotional film collection. The film documents a land consolidation project in the Maas en Waal-West area, located on the south bank of the Waal, inland from the Delta, across the river from Dodewaard. During this period, just after World War II, one of the first integrated rural development projects was undertaken here, parallel to the post-1953 flood reconstruction efforts in the southwestern Netherlands.

The film is structured in a format reminiscent of a Hollywood feature film. The protagonist, Van den Akker (Figure 6), a peasant, struggles under the weight of his burdens while dreaming of a better future. By the end of the narrative, Van den Akker is a content farmer, fully committed to the modernization of agriculture and rural life. A copy of this production with English subtitles is available in the Framing the Dutch Landscape series playlist on YouTube (Veer, 2024).

In this analytical approach, a film is treated as a rhetorical artifact, part of a broader societal discourse. The sender (filmmaker or institution) employs rhetorical techniques to persuade the receiver (viewer). The key components of this analysis include its claims and its warrants. Claims are defined as the film's core arguments, often embedded in the spoken narrative (voiceover). The warrants offer the justification for these claims, often presented visually rather than through abstract verbal statements. In *Van oud naar nieuw*, visual metaphors embedded in the real world function as warrants, reinforcing the film's ideological message.

In cultural studies, key concepts serve as analytical tools to structure and interpret meaning in texts, images and films. Common key concepts in film studies include problems, dominant values, wishes and dreams. Each research project may require specific additional key concepts derived from historical and cultural contexts. For this analysis, inspiration was drawn from the work of Hofstee (1960). In the same period that *Van oud naar nieuw* was produced, Hofstee formulated his sociographical theory on modern and traditional cultural patterns. His hypothesis suggested that to overcome the socio-economic stagnation of traditional rural life, the countryside and its inhabitants must be transformed through carefully planned development projects – turning peasants into efficient farmers and ideal rural citizens in well-structured, modern environments (Hofstee, 1960; see Van den Ban, 1982).

The key concepts in this context correspond to the rhetorical structure of the film. To locate key sequences, one must analyze the voiceover in relation to these key concepts. Similar to close reading in literary analysis, sequences where the spoken text explicitly references to key concepts constitute the core argument of the film. In a newly created 'inventory film', the most meaningful rhetorical sequences of *Van oud naar nieuw* were compiled. This inventory film, a short audiovisual work, is titled AV Rhetoric Analysis, Inventory Film 2, Representation 'Landscape' 1957; it is available in the Framing the Dutch Landscape playlist (Veer, 2024). For the theme 'landscape', these sequences were arranged chronologically.

The cinematographic inventory, and thus the film, opens with traditional landscapes, depicting the harsh working conditions and poor living standards of Van den Akker and his family. These visuals establish the traditional cultural pattern, emphasizing the restrictive nature of the old landscape, holding back the entrepreneurial spirit. According to agricultural engineers, the outdated landscape was economically inefficient, needing modernization. The land consolidation program is metaphorically presented as a 'rejuvenation treatment' for the region. The modernized landscape in the final sequences represents the desired landscape, where small farmers' problems are resolved, regional poverty is eradicated and social backwardness is eliminated. The new design mirrors the Northeast Polder, a well known example of large-scale land reclamation. The film also subtly promotes an 'American Rural Dream' – a mechanized, large-scale agrarian model. The film concludes in the rejuvenated area, with Van den Akker settled in a modernized farm, symbolizing the realization of Hofstee's theory.

The Ministry of Agriculture was proud of the results in the Maas en Waal-West area and made it a showcase for future land consolidation projects in the Dutch countryside nationwide. The aim of the film was to inform farmers of the benefits of the programme. The production was shown at information meetings about land consolidation projects all over the country, which explains why geographical names in the film are absent in the film. In the Ministry of Agriculture's overview of subjects in the film catalogue for 1960, *Van oud naar nieuw* is mentioned in six different categories. This indicates that the film's story was in line with the dominant narrative of the integrated rural development approach of the land consolidation policy. *Van oud naar nieuw* represents the ideologically charged narrative of the land consolidation programme and rural modernization policy.

This audiovisual rhetorical analysis shows that the main message of the film is that the enormous state interventions in landscape and society are in the interest of the individual farmers. The production metaphorically portrays the mental dimension of the ideologically tinted interventions as a farmer's personal dream. The landowners, the farmers, held agency over the project in their own area by law. And, as with so many regional projects on a national scale, the farmers also had agency over the success of state policies for rural development. So, to conduct the land consolidation program, the Ministry of Agriculture had to influence the farmers in order to create a positive attitude to the land consolidation projects. That is why the film ends with Van den Akker and his family are happily living and working on the modern farm.

Whether the production was successful in instilling a positive attitude towards the projects among farmers is difficult to assess. For more than twenty years the film, with the same description and without any comments, was part of the active film distribution collection of the Ministry. However, the Ministry of Agriculture's 1982 catalogue labels the production as 'not available'. Later that year, the film *Alles wisselt* (*Everything Changes*; Goudsmit, 1982) was released to inform the public about the regional projects conducted under the upcoming 1984 Rural Land Development Act.

Conclusion: Don't Get Lost in Translation

Scientists in the field of urbanism, engineering and water management can gain cultural-historical insight by exploring historical Dutch documentary films. This approach brings together the paradigms of the technical sciences and the humanities, creating a journey that is both inspiring and creatively challenging, albeit sometimes a bit perplexing (Bod, 2015). The final section of this essay offers suggestions from the perspective of film studies for researchers in technical disciplines. An important insight from the paradigm of cultural studies is that, just like other cultural products, an informative film is a part in a societal exchange of ideas, a debate full of rhetoric techniques, cultural values and ideological assumptions. When it is an archival film, it represents issues in a given region in times gone by. So, the study of historic films can be of great value to give insight into the dominant discourse. It's difficult, however, to find a proper science-based method to study the enormous amount and unequal types of data in films. The methods and outcomes are always open for debate. The assignment of a researcher is not to find the 'the truth' on a topic, nor to design technical solutions to problems in society, but, based on facts and using proper arguments, to give better insights into the researched themes in a reliable and trustworthy way.

In his study Henri Lefebvre introduces the craft of the rhythmanalyst. He argues that rhythmanalyst 'will have some points in common with the psychoanalyst' (1992, 19). I suggest to borrow elements from the anthropological paradigm. This discipline on the edges of social sciences and the humanities analyses human behaviour, cultures, social connections and languages. Film is an excellent medium to study these topics, because it can capture these practices in a very direct way. Thus, the genre of documentaries with an 'anthropological approach' can give clues as to the impact of the Delta Works, agricultural mechanization or other large-scale programs had on the physical, social and cultural domains. They can show scenes of daily life before, during and after the works. The representations in films of the routines in these three domains can give clues as to how local communities experience and value the changes in their social and ecological environments.

In film studies, all kinds of analysis methods can be found. They vary from deeply psychological or philosophical studies to technical inventories and statistical analyses. The last type of research focuses on frame size, the represented physical forms, used colours, spoken words or tones of music used in films or fragments of films. To find clues for the intentions the filmmaker wants to contribute to the societal debate, audiovisual rhetoric analysis can be of use. However, such an analysis is time-consuming. An alternative is the close viewing technique, as shown in the example of *Van oud naar nieuw*.

Most of the Dutch delta behind the dikes consists of rural land used for agriculture. The analytical video essays highlight that, in the Netherlands – much like the rest of western Europe – perspectives on rural areas are largely shaped by urban viewpoints (Woods, 2005; Bieleman, 2008). In this context, the impact of the Delta Works on agriculture remains an important but often overlooked consideration.

The first practical suggestion I would like to give as a film historian is to describe various genres in the corpus of audiovisual productions. In this essay two genres emerged. The first is 'engineering films'. They can be useful to understand the historical developments of paradigms in water management, urban design and engineering. However, they provide hardly any insight into the impact of such projects on local people. Their cinematographic stories pay no attention to the impact of such works on the physical environment and ecology. These themes are outside of the paradigm of the films, so any negative impact is probably seen as collateral damage. The other genre mentioned in this essay is the 'anthropological film'. In line with the intentions of this book, these films are, in my opinion, more suitable sources of information.

To improve the study of documentaries in the technical disciplines it would be wise to find information about the context of the productions. To get a first impression of the intended contribution to a debate, one could take a closer look at the opening and closing titles of the production. Usually, lots of metadata are given in written text in the audio and visual lines of the opening scenes and in the last resolution sequences at the end of the storyline. There, hints can be found about the aim of a film, information about the commissioner, the filmmaker and the audience they had in mind. Also, some indications about the aim, message and location may be found, and the societal debate to which the production aimed to contribute.

Analyzing films is not 'hard science'. Audiovisuals are multi-interpretable; this can be very exciting, but also confusing for scientists. There is always a chance to get lost in interpretation and translation from one paradigm to the other, from one's own cultural bubble to another, from the language of film to written text. In his study on space, time and everyday life, Lefebvre has practical advice to avoid getting lost: always go back to the physical rhythm of one's own body and look with empathy to the people in the local situation in any given period of time. I add the suggestion to the film researcher or rhythmanalyst to question oneself: how would I experience and value the situation, if I was in the position of a character in the represented story? This advice could also be very useful for all people who get lost in the worlds of theories, models and animated realities.

Figure 1: De Waalbandijk at Dodewaard along the river Waal on 31 May 2024. Photo: Peter Veer.

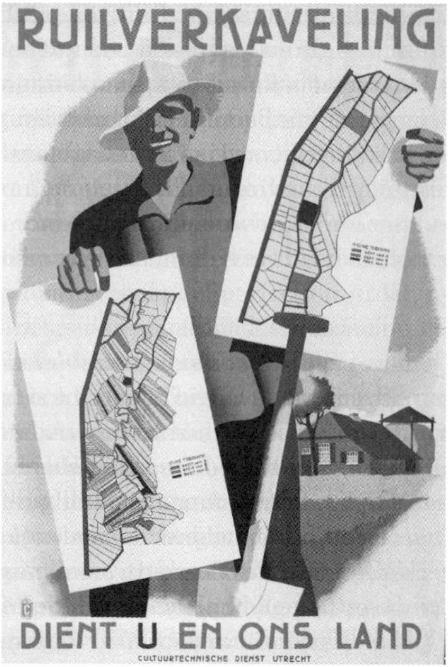

Figure 2: A poster of the Ministry of Agriculture to convince farmers of the usefulness and necessity of land consolidation. Cultuurtechnische Dienst Utrecht, 1946. VIII-1955-0402, Stadsarchief Rotterdam (Public domain).

Figure 3: An anthropological-like sequence. A farmer and his sons evaluate the large-scale land consolidation works. Still from *Eiland van vertrouwen* (Fernhout, 1950). © No further use allowed.

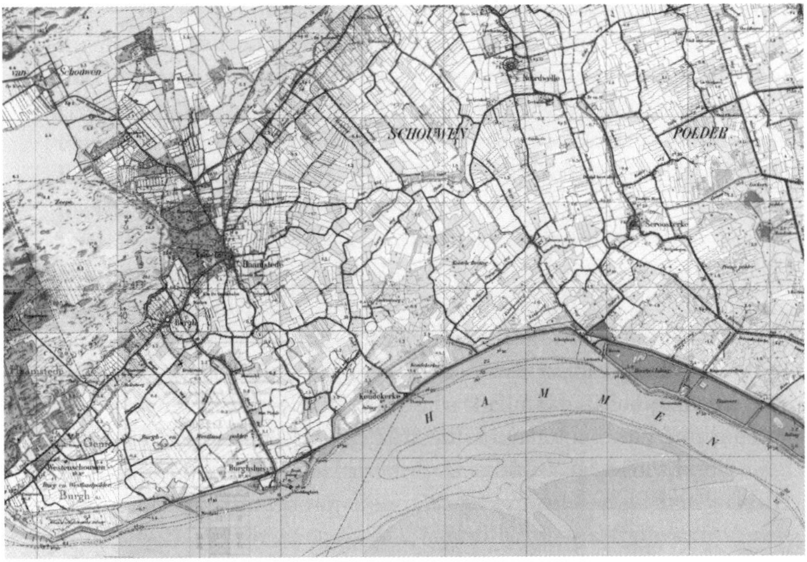

Figure 4: Schelphoek, Schouwen, 1952, before the floods. Kadaster, Apeldoorn. Topotijdreis, 2024.

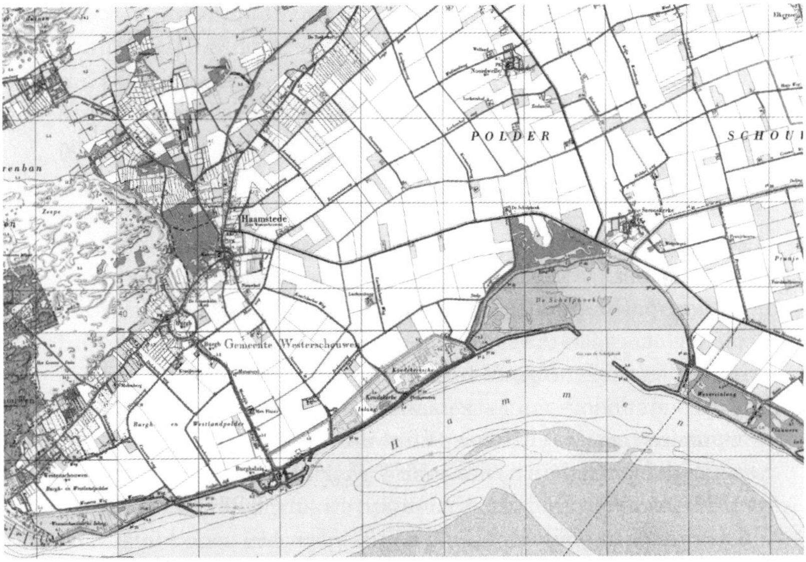

Figure 5: Schelphoek, Schouwen, 1962, after land consolidation. Kadaster, Apeldoorn. Topotijdreis, 2024.

Figure 6: Still image from *Van oud naar nieuw* (1957). © No further use allowed.

References

Acland, C. R., & Wasson, H. (eds., 2011). *Useful Cinema* (Durham, NC: Duke University Press).

Andela, G. (2000). *Kneedbaar landschap, kneedbaar volk* (Bussum: Thoth).

Banks, M. (2001). *Visual Methods in Social Research* (London: Sage).

Bieleman, J. (2008). *Boeren in Nederland: Geschiedenis van de landbouw 1500-2000* (Amsterdam: Boom). Abbreviated English version: Bieleman, J. (2010). *Five centuries of farming. A short history of Dutch agriculture 1500-2000* (Wageningen: Wageningen Academic Publishers).

Blom, A. (ed, 2017). *Post-War Reconstruction the Netherlands 1945-1965: The Future of a Bright and Brutal Heritage* (Rotterdam: NAi Publishers).

Bod, R. (2015). *New History of the Humanities* (transl. L. Richards). (Oxford: Oxford University Press).

Boelens, J, (director, 1975) *Als het water stijgt* [Film].

Davids, K., & 't Hart, M. (eds., 2023). *The World and the Netherlands: A Global History from a Dutch Perspective* (transl. V. Collingwood). (London: Bloomsbury Academic).

De Haas, M. (2013). *Two Centuries of State Involvement in the Dutch Agro Sector* (The Hague: Netherlands Scientific Council for Government Policy). Available: https://english.wrr.nl/publications/publications/2013/11/04/two-centuries-of-state-involvement-in-the-dutch-agro-sector (accessed 2 November 2024).

Eisenstein, S. (director, 1929). *The General Line aka The Old and the New* [Film].

Ellwood, D. (1989). *The Marshall Plan Forty Years After: Lessons for the International System Today* (Bologna: The Bologna center of the Johns Hopkins University School of Advanced International Studies).

Fernhout, J. (director, 1950). *Eiland van vertrouwen / Island of Trust* [Film]. John Ferno. Available: https://youtu.be/9rV1S1YAPJM (accessed 2 November 2024).

Flaherty, R. J. (director, 1922). *Nanook of the North* [Film]. Available: https://www.youtube.com/watch?v=3IAcRjBq93Y (accessed 9 November 2024).

Foucault, M. (1975). *Surveiller et punir: Naissance de la prison* (Paris: Gallimard). English translation (1977): *Discipline and Punish: The Birth of the Prison* (transl. A. Sheridan). (New York, NJ: Vintage Books).

Goudsmit*, Joan*. (director, 1982). *Alles Wisselt* [Film]. Carillon Films. Available: https://www.youtube.com/watch?v=HPDijiXIUgk&list=PL8vOt6SSgJj8iLMRMilmdMOzwMLOCCjZG&index=6 (accessed 9 November 2024).

Grierson, J. (director, 1929). *Drifters* [Film]. Available: https://www.youtube.com/watch?v=bqeHMh7ViLg (accessed 9 November 2024).

Hall, S. (ed., 1997). *Representation: Cultural Representations and Signifying Practices* (London: Sage).

Hediger, V., & Vondereau, P. (eds., 2009). *Films that Work: Industrial Film and the Productivity of Media* (Amsterdam: Amsterdam University Press).

Hediger, V., Hoof, F., & Zimmermann, Y. (eds., 2024). *Films That Work Harder: The Circulation of Industrial Film* (Amsterdam: Amsterdam University Press).

Hesling, W., & Peters, J.M. (1985). *Audiovisuele retoriek* (Leuven: Centrum voor Communicatiewetenschappen).

Hofstee, E. W. (1960). *Fundamenteel sociologisch speurwerk in het kader van het landbouwkundig wetenschappelijk onderzoek* (Wageningen: Landbouwhogeschool).

Karel, E. H. (2005). *De maakbare boer: Streekverbetering als instrument van het Nederlandse landbouwbeleid* (PhD thesis, University of Groningen/Wageningen N.H.A.I.).

Lefebvre, H. (1992). *Éléments de rythmanalyse (*Paris: Éditions Syllepse). English translation (1992): *Rhythmanalysis: Space, Time and Everyday Life* (transl. S. Elden & G. Moore) (New York: Continuum).

Lumière, A. M. L. N., & Lumière, L. J. (directors, 1895). *L'arrivée d'un train en gare de La Ciotat* [Film]. Available: https://www.youtube.com/watch?v=1FAj9fJQRZA (accessed 7 April 2025).

Marsman, H. (1941) [1936]. 'Herinnering aan Holland' [poem], in: *Verzamelde gedichten* (Amsterdam: Querido). English translation (2006). 'Memory of Holland' (transl. P. Vincent). Available: http://www.vlinder-01.dds.nl/cdr/poetry/memory_of_holland.htm

Mijn Gelderland (2024). *Verbeelding van de Waal* [website]. Available: https://mijngelderland.nl/specials/verbeelding-van-de-waal (accessed 10 October 2024).

Ministerie van Landbouw en Visserij (1960). *Films* [Catalogue]. Available: https://files.beeldengeluid.nl/pdf/Bewogen-landschap_bijlageA_catalogi.pdf (accessed 7 April 2025).

Ministerie van Landbouw en Visserij (1982). *Films* [Catalogue]. Available: https://files.beeldengeluid.nl/pdf/Bewogen-landschap_bijlageA_catalogi.pdf (accessed 7 April 2025).

Monaco, J. (2009). *How to read a film,* 4th ed. (New York: Oxford University Press).

Nichols, B. (2010). *Introduction to Documentary* (Bloomington, IN: Indiana University Press).

NOS (2020). *Hoog water en dreigende doorbraken in 1995* [Film compilation of historic news reels]. Available: https://nos.nl/video/2320560-hoog-water-en-dreigende-dijkdoorbraken-in-1995 (accessed 9 November 2024).

Penning, A. (director, 1957). *Van oud naar nieuw/From the Old to the New* [Film]. Haarlem; NV Multifilm, Available: (Dutch) https://www.youtube.com/watch?v=lw3Cvzv4v_o&list=PL8vOt6SSgJj8iLMRMilmdMOzwMLOCCjZG&index=2 and (English subs) https://youtu.be/nwVBdIsNhrY (accessed 7 April 2025).

Reporting the Delta (2024). *Reporting the Delta* [Instagram]. Available: https://www.instagram.com/reportingthedelta/ (accessed 10 November 2024).

Rogers, E. M. (2003). *Diffusion of Innovations*, 5th ed. (New York: Simon & Schuster).

Rose, G. (2012). *Visual Methodologies* (London: Sage).

Tennessee Valley Authority (1944). *The Valley of the Tennessee* [Film]. Office of War Information. Available: https://www.youtube.com/watch?v=SfJt-W7fwTY (accessed 9 November 2024).

Topotijdreis (2024). *Kadaster Topotijdreis* [website]. Available: https://www.topotijdreis.nl/kaart/1952/@43766,413026,9 (accessed 2 November 2024).

Van den Ban, A.W. (1982). *Inleiding tot de voorlichtingskunde* (Meppel: Boom, 1982).

Van Essen, J. (director, 1956). Luctor et Emergo [Film]. Amsterdam, Mirror Film. Available: https://www.beeldengeluid.nl/kennis/projecten/bewogen-landschap/landbouw-films-en-verhalen (accessed 10 November 2024).

Van Leeuwen, Th. (2005). *Introducing Social Semiotics* (London/New York: Routledge).

Veer, P. (2020a). *Bewogen landschap: Een cultuurhistorische studie over de filmpraktijken van het ministerie van Landbouw 1945-1985* (University of Amsterdam; Hilversum: Nederlands Instituut voor Beeld en Geluid. [PhD Thesis]. Available: https://dare.uva.nl/search?identifier=739504a9-2840-42b8-89e4-212fa3f1df53 (accessed 10 November 2024).

Veer, P. (2020b). *Bewogen landschap: Een cultuurhistorische studie over de filmpraktijken van het ministerie van Landbouw 1945-1985* (University of Amsterdam; Hilversum: Nederlands Instituut voor Beeld en Geluid. [Audio-visuals]. Available: https://beeldengeluid.nl/kennis/projecten/bewogen-landschap (accessed 9 November 2024).

Veer, P. (2024). *Framing the Dutch Landscape* [series playlist]. Available: https://www.youtube.com/playlist?list=PL8vOt6SSgJj_LRNVmQvEze_kWGJg0T9P8 (accessed 9 November 2024).

Vos, C. (2004.) *Bewegend verleden: Inleiding in de analyse van films en televisieprogramma's* (Amsterdam: Boom).

Waterschap Rivierenland (n.d.). *Dijkversterking Neder-Betuwe* [website]. Available: https://www.dijkversterkingnederbetuwe.nl/default.aspx (accessed 11 November 2024).

Woods, M. (2005). *Rural Geography* (London: Sage).
World Bank (2024). *Agricultural Land (% of Land Area) – European Union.* Available: https://data.worldbank.org/indicator/AG.LND.AGRI.ZS?-locations=EU&name (accessed 11 November 2024).

Research-by-Filmmaking in the Built Environment

Saman Seyff

Filming has commonly been used as an assistive method for data collection in architectural research; however, filmmaking is rarely recognized as a research method in architecture and urbanism. Moreover, despite its potential, architectural film is hardly accredited as an academic research outcome. This essay explores architectural filmmaking as a versatile research method that captures the dynamic essence of spatial characteristics while communicating scientific arguments through creative narration. It addresses filmmaking as a creative platform for investigative mapping, curating archives, and storytelling that, by connecting dots, reveals hidden layers of space-time and adds depth to scientific discourse. Architectural filmmaking, and the architectural film (ArchFilm) as its outcome, offers both a process and a medium that enable researchers to construct scientific arguments with artistic aptitude, enriching research by combining analytical rigor with imaginative visualization. This approach expands research dissemination, allowing ArchFilm to reach a broader audience beyond academia and foster public engagement. The essay also addresses challenges in employing filmmaking in academic contexts, such as maintaining objectivity and scientific accuracy. It advocates for a dynamic, interdisciplinary methodology that bridges science and art in perceiving and presenting the built environment.

Welcome to the Club!

Every time I attend presentations at architecture conferences or events, I'm struck by how often valuable film material is presented in a surprisingly unsophisticated way. Typically, a long video file is opened, and the presenter struggles to locate and play a short clip. This often involves dragging the timeline back and forth, rewinding and fast-forwarding, sometimes with accompanying sound issues. Ironically, these motion picture materials are often visually compelling and deeply relevant to the research being presented. Yet, rather than embedding selected clips directly into the narrative (archival or newly filmed) many presenters leave the audience watching a technical struggle.

If the act of navigating the film isn't central to the research, why not extract and edit the relevant segments in advance? After all, when quoting a book during a talk, a presenter doesn't flip through the entire volume on stage; they simply cite the relevant excerpt with proper context and attribution. So why isn't this same logic applied to film? It's unlikely to be a lack of technical ability; basic video editing today is as accessible as word processing. The issue, rather, seems to lie in a failure to fully integrate moving images into the research narrative with the same care and attention given to textual references.

This problem becomes even more clear when multiple films are involved. Here, I refer again to the most conventional academic writing approach, where scholars typically develop a thesis within a structured argument, supported by evidence-based analysis and proper citations from diverse references, all embedded in an original narrative. This approach can be equally applied to motion picture materials.

This essay does not aim to critique researchers' presentation preferences; rather, it focuses on exploring how the inherent potential of architecture can be approached to push the boundaries of traditional research and enrich our understanding of the built environment. The emphasis, therefore, is on leveraging this potential for the sake of research and acknowledging and validating it within scholarly inquiry. This methodological note aims to explore architectural filmmaking as a legitimate and creative research method.

Applying filmmaking as a research method was, for me, the result of a process. It began with producing video montages of archival materials to explain the context. My research particularly focuses on investigating the post-World War Two transformation of cities through the lens of their leisurescape, with a spotlight on clubs; it explores clubs not as singular and isolated buildings, but rather as socio-political and spatial assemblages within the urban landscape. As expected, buildings and urban fabric are mostly no longer the same today as they were in the 1960s and 1970s. Moreover, in the case of my research, at the time of their construction, many of the clubs were developed with little or no blueprints or other design and construction documents. Consequently, there is usually no direct set of archives accessible, so I had to curate a set of archives from indirect sources such as documentaries, (mainstream) fiction films, oral histories, podcasts, memoirs, photo albums, catalogues, media reports, novels and, of course, maps.

A simple question, therefore, arose: how can this archipelago of archival materials, in various formats such as video, voice, text, photographs and cartographs, be used both in the research process and for presentation? Would traditional methods, such as mapping through 2D cartography, effectively contribute to unveiling hidden layers? As usual, finding a persuasive answer to a simple question is not easy. After a series of examinations and considering feedback from academic and public audiences through several presentations, my conclusion was filmmaking – a method that remains undervalued or passively consumed within the academic sphere.

It seems valid to argue that the complex landscapes of our living spaces across the globe, characterized by multilayered and overlapping meshes, suggest that research in architecture and urbanism can rarely be adequately conducted using only a single quantitative or qualitative method. Accordingly, scholars tailor a set of mixed methods to meet the specific requirements of each research project. On the other hand, presenting scientific findings, even to an academic audience, poses a real challenge. It is in this regard that, in many cases, the outcomes of research may be uninspiring and, I would argue, insufficient and passive, as though we have overlooked the tools that the spirit of architecture has inherently provided us. By 'architecture', in this essay, I refer to the term in its broadest sense, encompassing everything from interior spaces to metropolitan areas.

Architectural Cinematics

Architecture, in its very DNA, engages with space-time and the events that unfold within; film, in another way, operates on the manipulation of space, time and movement. This similarity is not merely metaphorical; it is embedded in how architecture and cinema both engage the body and senses to produce a temporal flow through space.

The mutual fascination of architecture and cinema, as Beatriz Colomina (1994) asserts, often blurs the lines between interior and exterior, private and public, reality and illusion. Film, according to Colomina, does not just capture architecture; it actively participates in its interpretation and meaning-making process. The cinematic language, then, according to Giuliana Bruno (2002), can articulate the experiential qualities of architecture, capturing the sensory and temporal dimensions of spaces that traditional architectural drawings and photographs might miss. In this respect, architecture in film becomes a medium through which the filmmaker can explore broader social and cultural narratives.

The qualities rooted in the shared attributes of cinema and architecture essentially allow filmmaking to be considered an active, process-driven research method in the field of architecture and urbanism. This method, therefore, goes far beyond merely portraying architecture in film and addresses the dynamics and articulation of space, time and movement.

Through the synergy of these shared attributes and the inherent cinematic qualities of architecture – which I refer to as architectural cinematics – filmmaking as a research method enables researchers to investigate the intrinsic qualities of the built environment, emotional resonance and ideological critique, thereby creating dynamic architectural narratives. Although an individual building may appear cinemagenic, beyond the zoom-in and zoom-out of interior and exterior perspectives, this method creates space for explorative research into architecture as a socio-spatial assemblage within its context and in relation to the multilayered and overlapping meshes of the built environment.

In the book *Architecture Filmmaking* (2020), editors Igea Troiani and Hugh Campbell discuss filmmaking as a tool for research and explore how film can serve as a medium for 'representation and analysis', examining the spatial and temporal dimensions of architecture; 'critical reflection', facilitating new ways to analyze and reimagine architectural design and its societal implications; and 'pedagogical innovation', enriching architectural education through the visualization of movement, temporality and other dynamic elements. The volume includes contributions from architects and filmmakers worldwide, offering diverse perspectives on how film can inform architectural design, critique societal structures and enhance educational methodologies. Key sections address filmmaking techniques, the reimagining of architectural and urban spaces and the integration of film in architectural pedagogy, making it an essential resource for those interested in innovative approaches to architectural theory and practice (Troiani & Campbell, 2020).

Richard Koeck (2013), in *Cine-Scapes*, describes existing cinematic qualities in postmodern landscapes as a visual apparatus and a system that is based on movement, light and the body, which can be explored through kinematic, kinetic and kinaesthetic ways. He further elaborates on the idea of the urban environment as a cinematic space, where the rhythm of daily life, the flux of people and the play of light and shadow create a continuous filmic experience. This supports the idea that the city becomes a stage where the drama of everyday life unfolds and offers endless narratives to its inhabitants.

Addressing the architectural cinematics in the urban landscape, François Penz and Andong Lu, in their edited book *Urban Cinematics* (2011), approach the topic methodologically in order to understand urban phenomena through the moving image. Cinema, according to them, has the unique ability to capture the fluidity and dynamism of urban life and provide an unparalleled perspective on the complexity of cities. Furthermore, cinema not only reflects urban spaces but also actively participates in the production of urban imaginaries. Penz and Lu argue that through the lens of cinema, the city becomes a site of endless negotiation between reality and representation.

Researching Otherwise

The arguments presented in this essay are generally applicable to any approach to filmmaking, whether through original filming, the montage of pre-existing archival materials or a combination of both. This validity stems from the common values inherent in all forms of filmmaking, which methodologically empower the researcher-filmmaker to 'research otherwise'.

I borrowed the title of this section from the book *Researching Otherwise: Pluriversal Methodologies for Landscape and Urban Studies* (2024), in which filmmaking is one of the highlighted methods. In this book, Nitin Bathla, the editor, refers to his collaboration in producing the documentary *Not Just Roads* (2021) and draws on David MacDougall's book *The Looking Machine* (2019), stating that the use of the cinematic method 'does not become a way of narrating something that he already knew as a researcher, but rather centres on the cinematic imagination' (Bathla, 2024, 30). Through the cinematic imagination, the researcher reveals hidden realities within a spatial ecosystem and documents an interpretation that might differ from the official narratives of that context.

Functioning like a map, film orients the filmmaker-researcher and viewer within the narrative of the space-time, while simultaneously reshaping their understanding of the living environment. Tom Conley (2007), in *Cartographic Cinema*, argues that the filmic image itself constitutes an active process of mapping that translates geographic and cultural environments into visual forms. According to him, cinema and cartography both provide ways to navigate and comprehend spatial complexity, linking the visual field with territorial meaning. Conley investigates how films can act as tools of spatial exploration, mapping both real and imagined terrains, and how films operate as cinematic maps that mediate our relationship to space, place and narrative.

Whether filmmaking is done through archival footage montage or original filming, the moment a film is produced it holds the potential to become archival material. However, how uncut materials are initially produced and the type of footage which is used essentially influences the researcher's approach to filmmaking and the research process, which in turn alters the methodological framework. In the case of my research on leisurescapes and clubs, the entire methodological framework is shaped around pre-existing reference materials. Consequently, filming and recording original footage are not involved; instead, the emphasis is on montage and creating a collage that incorporates time as a fourth dimension.

The use of filmic techniques, such as montage and juxtaposition, as Juhani Pallasmaa (2008) suggests, can reveal the relational qualities between different architectural elements and their contexts. Additionally, Shiel and Fitzmaurice (2001) argue that audio-visual archives, when combined with narrative film techniques, can convey complex historical and spatial relationships in an accessible and engaging manner. Through the process of selection, interpretation and presentation, incorporating filmmaking into archival research offers a dynamic approach to historiography. It allows the researcher to investigatively map the unfolding life of the space as it was captured at the time and produce a distinct narrative derived solely from the montage of archival materials. This approach thereby positions the researcher-filmmaker, by default, as both a producer and curator of the archive, enhances the communicative power of architectural research, and fosters a deeper understanding of the historical spirit of the built environment.

On the other hand, when original filming through the fieldwork is the initiative, Bathla (2024, 29) explains that 'recording, regardless of the media we choose' and 'similarly, editing of media – be it film, image, sound, drawing or text – forces the researcher to select what is meaningful to a specific situation or place, . . . [which] is a form of analysing'. In the process of filmmaking, Bathla (2024, 30) acknowledges 'recording, as in paying attention; editing, as in analysing; and then later narrating everything to the viewer as a set of sequences'. Emphasizing filming, Denise Bertschi (2024, 216) points out 'investigative filmmaking' as a tool for mapping to be 'researched otherwise'. Bertschi refers to Tim Ingold's (2022, 148) theory of 'place making through movement' and suggests 'mapping through walking'. She describes this phase as 'a performative act' and 'a gesture of the body', based on Edward Soja's thesis that 'space matters' (Bertschi, 2024, 217). Through walking with the camera, 'the researcher physically emplaces themself in the urban space which they wish to investigate'. Through this method, by moving 'between buildings' and 'around and within', these architectural objects 'become the sites of investigation, where the researchers' primary sensory experiences are made' (*ibid.*).

Situating investigative filmmaking in contrast with classical cartography, Bertschi (*ibid.*), in addition to filming, highlights editing and subsequently presenting as integral components of the process in her research: 'through editing the filmic sequences ... the film suggests a potential range of new relations between these places, assembled through cutting and rearranging'. Referring to Ingold's interpretation of the 'meshwork' of paths with varying intensities (Ingold, 2007, 80) and Henri Lefebvre's notion of the meshwork, by which he means the reticular patterns left by people (Lefebvre, 1991, 117-118), she further explains that 'the choice to present the final film as a multi-screen, with three different simultaneous projections, further highlights the interweaving of these places into one net of relations, or meshwork' (Bertschi, 2024, 217-218).

It Is All About Process

'Process' is the key concept that enables architectural cinematics to position filmmaking as a suitable method for architectural research. More precisely, filmmaking is a matrix of processes in each phase, in which the components, although they could be deployed individually, resonate in interaction to each other. The key point is that we are talking about a creative, exploratory, process-driven method where the researcher, as the filmmaker, can utilize all or some of these components in a manner that is both suitably and effectively tailored for the research. Just like 'research by design' or 'research through design', it is 'research by filmmaking' or 'research through filmmaking'. Similar to research through design, where the design process serves as a central tool for investigation, research through filmmaking constitutes a methodical approach, where filmmaking itself becomes an active form of inquiry. Just as research through design engages creative practices of design to uncover insights, research through filmmaking uses architectural cinematics to explore complex phenomena. This method fosters critical understanding by employing the act of filmmaking – through composition, editing and presentation – as a means to engage with, analyze and interpret phenomena. Therefore, the process of making ArchFilm becomes an integral part of knowledge production in architecture.

A simplified, universal set of phases, providing a schematic overview of filmmaking, are Pre-Production, Production, Post-Production and Exhibition, consisting of the components listed in Table 1.

Phase	Component
Pre-production	Scriptwriting Storyboarding Casting Location scouting
Production	Directing Cinematography Acting Sound recording
Post-production	Editing Sound design & mixing Visual effects Rendering
Exhibition	

Table 1: Standard filmmaking phases and components overview 1. Table compiled by Saman Seyff, 2024.

This universal phasing also applies to architectural filmmaking, particularly when used for fictional storytelling. However, even when ArchFilms are considered non-fiction or documentary, they remain a production and, like any other, involve pre-production, production and post-production phases.

To be more precise, in Table 2, I suggest a tailored phasing for non-fiction architecture filmmaking, which is more flexible and detailed than the universal standard phasing but can still, like the standard, be executed non-linearly. For instance, one could start by script writing and forming the narrative, then proceed to filming and preparing audiovisual materials, while another filmmaker might approach the process the other way around.

Alternatively, in cases where the main research materials are pre-existing archival, as in my research, the editing phase becomes more prominent, as if footage montage is a tool for archival analysis. Hence, there is no first-in/first-out stack and no clear boundary between phases. When the boundaries become blurry, writing, editing and analyzing form a cyclical dynamic that assists the researcher in creating a customized process or, rather, a tailored method derived from filmmaking phases. This shows that phases and components do not necessarily need to be executed linearly or sequentially; instead, they can be approached in a back-and-forth, interchangeable process. Indeed, it is all about the process!

The phasing suggestion in Table 2, although it could serve as a guide or outline of instruction for some researchers, is primarily intended to demonstrate an example of a framework.

Phase	Component
Concept	Concept development Research hypothesis Narrative development
Script and structure	Outlines Scriptwriting Storyboarding Case studies
Cinematography	Filming Archival Selection
Visual aids	Illustrations Drawings Geospatial analytics Infographics and charts On-screen text 3D modelling VR / AR
Sonic	Audio Ambiance Background music Narration and voiceover
Editing	
Ethical and Legal considerations	Accuracy and fairness Permissions and rights
Final rendering	
Exhibition	

Table 2: Architectural filmmaking phases and components overview. Table compiled by Saman Seyff, 2024.

The fascination of filmmaking as a research method lies in its exploratory, adaptable nature and its capability for in-process creativity, which also makes it perfectly compatible with other qualitative and quantitative methods. This exploratory, process-oriented and process-driven characteristic is the initial distinguishing factor of using filmmaking as a research method in architecture and urbanism, which also differentiates it from conventional documentary filmmaking. Given that both approaches share techniques such as narrative construction, visual storytelling and the documentation of subjects, it is noteworthy that architectural filmmaking goes beyond mere technicalities and representation to actively engage with the built environment as both an object of study and a medium of expression.

In architectural research, filmmaking is essentially employed to investigate spatial features, as well as the interactions and relationships with life as it flows through spaces. The focus, therefore, is not just on documenting facts but on interpreting the space through the researcher's lens. This reflective method expands the scope of how architecture is understood and communicated by encouraging researchers to not only present findings but also critically examine how spaces are perceived, experienced and transformed.

ArchFilm Negotiations

Among the many features and capabilities of architectural filmmaking as a tool of research method, the ones that stand out most to me are those that create room at some point for negotiation within the research-production process and open up space for greater interaction between the researcher and the study, as well as between the ArchFilm as the outcome and the audience. These negotiations, in a mutual way, both drive and are driven by analysis, interpretation and presentation. I refer to these as ArchFilm negotiations: positionality, storytelling and openness to a public audience.

However, one might argue that while ArchFilm negotiations can enhance the appeal and accessibility of research findings, they also risk compromising the objectivity and rigour required in academic work. As a rule of thumb, like any other scientific outcome, all scientific and ethical standards must be upheld in the production of ArchFilms. Hence, researchers, regardless of the methods used, must adopt a framework that ensures the scientific validity of their narratives. In the production of ArchFilms, the filmmaker, as Lucy Suchman (2007) states, must navigate the tension between artistic expression and factual accuracy to ensure that their work remains both engaging and truthful. The artistic aspects of filmmaking should, therefore, function to enhance rather than undermine the scientific validity of the research.

Thoughtfulness and deliberateness are also apt notions for addressing the challenge of maintaining objectivity when employing filmmaking as a research method. This method equips the researcher with a robust toolkit for exploring positionality, which, to a great extent, corresponds with Donna Haraway's critique of objectivity (1988). Considering Haraway's argument that all knowledge is contingent, influenced by context and always situated within a network of relationships, filmmaking provides a toolkit for producing 'situated knowledge' in architecture and urbanism, which, according to Haraway, is a more accountable, collective, reflexive and politically aware form of objectivity.

When a researcher employs filmmaking as a method, they initially position themselves as a filmmaker. This means that the researcher acknowledges their agency in creating narrative(s), an agency derived from the active nature of the method. The researcher-filmmaker must therefore choose not only the research lenses but also the film frames, which, to a degree, will overlap and function interchangeably. Moreover, the researcher is in a position to determine the degree of zoom in each frame, the rhythm of the frames and the use of original sound or voiceover. It is up to the researcher to determine how the audience will be navigated from A to B to C in both space and time. Filmmaking offers researchers a unique platform to articulate scientific arguments through visual storytelling. This dual approach – combining rigorous analysis with imaginative visualization – enhances the communicative power of architectural research.

Whether fiction or non-fiction, what distinguishes a storytelling narrative from a non-storytelling one? A simple answer is that while both types of narratives might describe a series of events, a storytelling narrative actively shapes those events into a cohesive and purposeful story with clear progression. A storytelling narrative involves more than just the description of events – it incorporates elements such as character (for example, the protagonist could be architectural, such as buildings), plot, tension and resolution to engage the audience and convey a deeper meaning or message. By employing narrative techniques and creative visual storytelling, researchers can convey complex architectural and urban concepts in a manner that is both engaging and accessible.

The assertion that the general public can more easily engage with a storytelling narrative in the form of a film, as opposed to an academic journal paper or book, is widely accepted. While the primary audience for practical research is typically not the general public, the use of cinematic media allows researchers to extend their reach to a broader audience, including those outside academia. This broader dissemination fosters public understanding and engagement with architectural and urban studies by effectively bridging the gap between science and art.

Filmmaking as a research method and as a powerful tool for visualizing and interpreting complex spatial and social phenomena also aids in disseminating knowledge. It allows for a narrative-driven exploration of architectural and urban issues, making them more accessible and engaging. This approach offers a unique way to democratize knowledge, making architectural research more relevant to the public and providing a platform for diverse voices to contribute to discussions about the future of living spaces.

The opportunities presented by filmmaking in architectural research are substantial, as it raises public awareness of existing dilemmas in living spaces and future concerns. Furthermore, it encourages citizens to participate more actively in discussions and decision-making processes related to the built environment. This is especially relevant when filmmaking is integrated with other research methods, such as focus groups, and employed in collaborative contexts like co-creation or participatory design. By combining these approaches, researchers can create a more inclusive and dynamic dialogue with the public, encouraging a more participatory role in shaping the spaces in which they live.

Ending

Cinema is a complex and delicate art/industry with over a century of history, continually evolving – technologically, technically, narratively and in terms of content – and often regarded as state-of-the-art. Nevertheless, leveraging cinematic techniques and investigating through film production processes offer researchers an enhanced toolkit that can be employed as a research method in the field of architecture and urbanism: research through filmmaking.

A review of the limited references cited here reflects the long-standing discussion on this topic, which has been active for at least 30 years. However, the limited application and outcomes so far highlight how much filmmaking as a method has been overlooked. Compared to 30 years ago, technological advancements have significantly transformed the approach to producing motion pictures, making all four phases – pre-production, production, post-production and exhibition – more accessible than ever before. This is evidenced by the widespread use of social media, where a mobile phone enables anyone to film and edit outside the traditional studio setting, and display their work without the need for a cinema screen.

Overall, I do not believe that engaging with filmmaking as a research method requires a researcher to be a proficient filmmaker; they can, however, utilize the processes and phases of filmmaking effectively. Given that inter- and multi-disciplinary research is often a collaborative effort, the presence of a cinematographer or editor in an architectural research project should be no more unusual than the involvement of an AI specialist or programmer in the field of architecture. The key point is that this method, in and of itself, does not impose limitations on its application. It can be employed as a simple qualitative method or in combination with other quantitative and qualitative methods as part of a mixed-methods approach.

Understanding the built environment – past, present and future – requires a thorough grasp of its complex, interconnected nature. To deepen this understanding, we must continually enhance our toolkit by integrating tools from various disciplines and areas of expertise. An inter-, trans- and multi-disciplinary approach is crucial for expanding the methodological scope of our research, while continuously updating and refining the capabilities of a scientific and systematic exploration of the built environment.

This essay is a methodological note on Architectural Filmmaking as an exploratory, active and process-driven method. I believe that, just as sketching has developed its own identity as a unique architectural language, the same can happen for ArchFilm, with 'architectural frames' establishing their own path. This debate, of course, did not start with this essay, nor will it conclude with its publication, yet the question remains: to what extent is the academic and research community open to alternative methods, and how agile is it in acknowledging less conventional approaches?

References

Bathla, N. (2024). 'A Methodological Pluriverse: An Introduction', in N. Bathla (ed.), *Researching Otherwise: Pluriversal Methodologies for Landscape and Urban Studies*, 11-34 (Zürich: GTA Verlag).

Bertschi, B. (2024). 'Mapping Architectural Traces of the Swiss-South African Gold Trade in Johannesburg: Film as Investigative Spatial Research', in N. Bathla (ed.), *Researching otherwise: Pluriversal methodologies for landscape and urban studies* 215-239 (Zürich: GTA Verlag).

Bruno, G. (2002). *Atlas of Emotion: Journeys in Art, Architecture, and Film* (London: Verso).

Conley, T. (2007). *Cartographic Cinema* (Minneapolis, MN: University of Minnesota Press).

Colomina, B. (1994). *Privacy and Publicity: Modern Architecture as Mass Media* (Cambridge, MA: The MIT Press).

Haraway, D. (1988). 'Situated Knowledges: The Science Question in Feminism and the Privilege of Partial Perspective', *Feminist Studies, 14*(3), 575-599.

Ingold, T. (2022). 'Against Space: Place, Movement, Knowledge', in *Being Alive: Essays on Movement, Knowledge and Description*, 145-155 (London: Routledge).

Ingold, T. (2007). *Lines: A Brief History* (London: Routledge).

Koeck, R. (2013). *Cine-Scapes: Cinematic Spaces in Architecture and Cities* (London: Routledge).

Lefebvre, H. (1991). *The Production of Space* (transl. D. Nicholson-Smith). (Hoboken. NJ: Wiley-Blackwell).

MacDougall, D. (2019). *The Looking Machine: Essays on Cinema, Anthropology and Documentary Filmmaking* (Manchester: Manchester University Press).

Marks, L. U. (2000). *The Skin of the Film: Intercultural Cinema, Embodiment, and the Senses* (Durham, NC: Duke University Press).

Pallasmaa, J. (2008). *The Architecture of Image: Existential Space in Cinema* (Helsinki: Rakennustieto Publishing).

Penz, F., & Lu, A. (2011). *Urban Cinematics: Understanding Urban Phenomena through the Moving Image* (Bristol: Intellect).

Shiel, M., & Fitzmaurice, T. (Eds., 2001). *Cinema and the City: Film and Urban Societies in a Global Context* (Hoboken. NJ: Wiley-Blackwell).

Soja, E. W. (2010). *Seeking Spatial Justice* (Minneapolis, MN: University of Minnesota Press).

Suchman, L. A. (2007). *Human-Machine Reconfigurations: Plans and Situated Actions* (Cambridge: Cambridge University Press).

Troiani, I., & Campbell, H. (eds., 2020). *Architecture Filmmaking: Framing the Discourse and Conditions of Production in Architectural Practice and Education* (Bristol: Intellect).

Memories in the High Tide

Nicoletta Traversa and Giuseppe Ferrari

RI-PRESE memory keepers is a home movies archive based in Venice, Italy. The archive is important for the issues that are dear to the Reporting the Delta project: from the role of Venice as a symbol of the dissolving of private memories, to the need of a process of preserving and sharing territorial heritage contained in the home movies – an unparalleled legacy of the community – to the issue of preservation and salvaging materials that otherwise run the risk of degradation due to the high tides. So, high water becomes a possible interpretation of the entire archive: the city of Venice has undergone countless events of flooding, rising steadily throughout the twentieth century, also due to climate change. The waters have contributed to shape, change and at times dissolve or suggest, stimulate or enhance new observation points on the landscape and on Venice's private and community memories.

Venice lagoon: the bow of a *sanpierota*, a wooden rowing boat with a flat keel, among the most popular in the Venetian lagoon tradition, is moving slowly forward. Reverse shot: a man, standing, is slowly rowing, watching the scenery. Ramo del Vino, San Marco district: a girl is walking back and forth in high water, along the *fondamenta* just outside her house. At the Zattere, a man and a woman are holding their child's hands tightly, watching over her first steps.

These people are living in the 1950s, the 60s, the 70s, and the scenes come from moving images contained in home movies, which in the last ten years we have been trying to safeguard through the Venetian archive project RI-PRESE memory keepers. RI-PRESE is a project of collection, safeguarding, digitization, enhancement and research, fully devoted to the audiovisual private memories of home movies. These are small-gauge, amateur movies shot since the 1920s, on film supports like 9.5 mm, 8 mm, 16 mm, super8 or on videotapes like VHS or video8, for non-professional purposes. Many show, for example, the birth of a child, weddings, sports and social events, community get-togethers, trips . . .

The project originated in 2013, inspired by a course we took at the Faculty of Design and Arts of the Iuav University of Venice, which led us to deepen our interest in the issues of safeguarding private memories and in the creative reuse of found footage for artistic and cinematographic projects. Between 2013 and 2014 we began to collect various materials, including projectors, cameras, video recorders and reels of amateur films and video cassettes. We tried to develop artisanal techniques for digitizing films, and we began to find out about their contents. We realized almost immediately the potential hidden in this audiovisual heritage, which forms precious memories for families, but is also an invaluable heritage of the communities which generated them. We started planning opportunities to enhance their contents, such as archive-based films, exhibitions and film concerts.

Inspired also by similar projects developed since the early 2000s in Italy and throughout Europe, we finally realized how essential it was to undertake the mission of founding a local archive, based in Venice, our adopted city. RI-PRESE was born in this way, conceived over years of experimentation, proceeding through trial and error. As the archive project grew, so did the number of people who became interested in our mission. Venetian citizens increasingly began to bring us their home movies, both to be able to review their contents, as well as to share our goal of spreading and protecting the collective memory of the city. Today RI-PRESE preserves more than 2000 individual home movies, shot on film between the 1920s and 1980s, to which must be added more than 700 hours of footage on videocassettes, shot from the 1980s to the end of the century.

The project is aimed at citizens of the Venice area, who are encouraged to join the territorial memory archive. This means that people can bring us their movies, and receive in turn a free high-quality digitalization, which allows them to reclaim their private memories. Often elderly amateur filmmakers themselves want to give their children and grandchildren their filmed memories. Or it is the descendants of amateur filmmakers who contact us: they have found a box full of films hidden in the attic, in the cellar, in some drawer at home. They have no idea how to recover the contents; often they have never seen the films, or they only vaguely remember their home screening.

Home movies tell the intimate stories of the people and families of Venice. Not only do they have a memory and sentimental value for every single one of them, but they are also crucial community heritage. These are pieces of a mosaic that can be placed close to one another, and the archive dynamics gives them the possibility to interact. So they can be seen in perspective, shared, researched and revised, and they make the story of the territory emerge, expressing a multifaceted, never exhaustive, never rhetorical narration. Those who add their films to the archive are always aware that these films, beyond 'going back home' in digital form, will become part of a collective project. It is important to take into account the social value of safeguarding and sharing the audiovisual memories that belong to the territory: who partakes in an audiovisual memory archive project like ours becomes an active contributor to the project. This key objective of RI-PRESE is also one of the motives that most captivates the people who join the archive, who feel like an active part of a project to revitalize collective memory. By sharing their own and their families' memories, they assume the responsibility to acknowledge that these memories also constitute the substance of an identity, for all the fellow citizens that recognize themselves in these images and memories. They thus contribute to recovering a sense of belonging to the community and consequently caring for it.

By being audiovisual entities, not fully structured, with holes, 'badly-made' – as highlighted by one of the most prestigious scholars of amateur cinema, Roger Odin – they are also ready to be used in many ways. Home movies set free creativity, suggest impressions and allow the subjectivity of film directors, artists and curators to put images and narrations back in a circle. They tend to enhance cultural projects also beyond a sheer documentary perspective (Odin, 1995).

The films in the archive find numerous opportunities of enhancement: they become part of film productions and museum exhibition projects, accompany concerts, and are studied and explored on occasions such as the Home Movie Days, a festival dedicated to private film memory which took place in Venice in 2023 in collaboration between the archive and the national association Re-framing home movies. Over the years, RI-PRESE has in fact developed a network of collaboration with institutions in the Venetian area and throughout Italy, in order to develop shared projects. Finally, at the university, but also in schools of the Venetian area, RI-PRESE carries out educational programs, dedicated to the discovery and reuse of the territorial audiovisual heritage stored in home movies.

Venice is a symbol of a place that feels an urge to not lose the roots of its recent past and to keep track of the city's memory, even more so in times like now, when the social fabric of the city itself is swiftly deteriorating. The depopulation of the historic city comes along with the depopulation of the memories the citizens recorded and preserved over the whole twentieth century. The research, collection, cataloguing, digitizing and enhancing operations that we implement in Venice through RI-PRESE are moreover considered urgent due to the nature of the materials involved. Home movies are recorded on supports exposed to a high risk of degradation, like films of cellulose acetate or magnetic tapes. As they are materials not traditionally thought of as objects to be preserved or archived outside the household, they are often kept in conditions that damage their state and compromise their materiality. The preservation time of home movies shot on acetate film is between 70 and 100 years and that of magnetic tapes about 30 years. But these evaluations are subject to conditions such as regulated temperature and humidity, which can barely be achieved in households, or worse, in private citizens' cellars or storerooms (Del Amo García, 2004). Moreover, unlike other audiovisuals like cinematographic, industrial or commercial films, home movies are usually sole copies. Losing them would mean to lose their content, the memory – irreparably.

Venice is peculiar also from the point-of-view of the climate. Struck by high waters with a growing frequency and intensity in the last decade, it is affected more than other places by the risk of a loss of memories contained on private, audiovisual supports, often kept and at times forgotten among other keepsakes left on ground floors. High water has touched the RI-PRESE archive at different levels: it is part of the story of the archive, is recorded in the Venetian home movies and is an agent potentially destroying and modifying film memories.

At 22.50 on 12 November 2019, due to a concurrence of climate and astronomical events (strong bora and scirocco winds, heavy rain and moon phase) the Venice lagoon was struck by the second-biggest high tide ever recorded since the beginning of the twentieth century. The resulting water height of 187 cm had not been forecast and caused massive damage to the city. This catastrophic high tide was called 'Acqua Granda', like a similar one which happened in 1966, and can be considered the real founding event for RI-PRESE, although an embryo of our archive had already existed for some years. That night, the high tide hit, along with hundreds of businesses, homes and rooms on ground floors of the historic city, also the place where we stored the materials and the first film stocks collected by the archive. Some movies ended up soaked by the salty water.

Back then, in addition to keeping the materials in a ground floor room, we also lived in a ground floor apartment. This event was therefore particularly traumatic for us. Discouraged and tired, we were unable to keep the situation under control. Some friends of the project, including other Italian archives, scholars and enthusiasts, suggested we use crowdfunding. This might allow us to cover the costs of restoring the films and the damaged materials. The 'help_riprese' crowdfunding encouraged us to open communication channels for RI-PRESE, such as social media pages, which did not yet exist (the project was based on word of mouth up to that point), and to try to spread our idea of a Venetian home movie archive as much as possible, introducing it officially to the city. The crowdfunding was successful: it received excellent media coverage and made the project known to many people who until then had not known about it. In a few months, through huge support, we managed to raise the necessary funds to repair the damaged films. They were washed and restored with the help of the *Archivio Nazionale del Film di Famiglia* in Bologna.

But the amount of donations, together with the help of a Dutch scanner manufacturer, *Filmfabriek*, and through personal investment, unexpectedly led RI-PRESE towards a new goal: to obtain a high-quality scanner for small-gauge films. This is a complex and high-performance machine, which allowed us, after years of efforts, to digitize films in a more professional way and at a significantly higher quality level. Being able to count on this type of equipment gave new life and courage to the project, allowing it to re-emerge from the drama of the flood, but also to better structure the operations, and to look with hope to the future of the archive itself. Furthermore, the trauma had allowed us as RI-PRESE founders to expand and enhance our awareness of how important and urgent the mission is to safeguard private memories, also in light of what had just happened and how many people had taken an interest in us.

High water, however, enters into the archive also in another shape: some home movies preserved by our archive portray the high tide as a matter of fact, making it the protagonist of the takes. This becomes one of the elements of the identity of Venetian home movies; it occurs in residents' movies as well as those shot by tourists who during the last century came to visit and portrayed the city. Actually, since its founding, we have strived on the one hand to preserve and enhance the memories of Venice's inhabitants, and on the other hand to bring images shot by tourists from all over the world back to where they were captured. This is done through research and acquisitions that the archive periodically carries out, but also through connections with non-Venetian archive entities. It was actually very common for amateur filmmakers who travelled to Italy or across Europe in the Twentieth Century to visit and film Venice. It is also very common for Venetian amateur filmmakers to dedicate part of their films to other cities or territories, given that vacation is one of the most common moments to shoot amateur films. RI-PRESE has been sketching out for several years the idea of a protocol that could bring together various local archives. In this way, memories would be shared with the territories they were 'extracted' from. Besides, orphan cinetouristic films that lost their connection with their family of origin and do not have information to trace them, may be reconnected with their makers. RI-PRESE tries to recover these orphan films and make them available to the archive, thus bringing these memories back to the city. All this is carried out in the belief that the cinetouristic images perform a gaze that has always been present and enduring in the city, capable of tracing trajectories and looks that are essential to take into account, and belonging to the identity of the city itself.

In both the movies of the 'residents' as well as in the 'cinetouristic' ones, high water is usually seen when it is 'harmless', not traumatic. It usually emerges when it is possible for the filmmakers to leave aside the worries about a dramatic climate event, leaving them free to record the change that the water causes to the spaces and the city landscapes. And so the masses of tourists in San Marco square give way to a big basin cut through by small waves and beneath the arches of the Procuratie Vecchie boats keep sailing. At Rialto, people let themselves be transported on a cart so as not to wet their feet. High tide redraws Venice, questions bias, enacts new visions. The water of the lagoon has the potential, right when it trespasses its borders and abandons its perimeters, to redesign and re-mediate the city.

There is also a third level at which the high tides of Venice come into the archive. Some film stocks have been affected by high waters likely several times over the course of the decades. This is the case, for instance, with the Gianese film archive. The collection is made up of 47 films in 8 mm format, shot from the 1950s to the early 70s, which had been kept in a box in a storeroom on the ground floor, forgotten or lost by the original family. The films were donated to our archive by someone who did not belong to the family and had found them by chance. No one knew the content of the reels. The film director Bill Morrison has become passionate about the RI-PRESE home movie archive. Through famous films such as *Decasia* (Morrison, 2002), *Dawson City: Frozen Time* (Morrison, 2016) and many others, he devoted his entire thirty-year career to the creation of cinematographic works about the decay of film supports and the relationship between destruction and new meanings. Starting from a meeting with Bill Morrison, an analysis was developed to find the identity of the original family through the films. This project, which at the time of writing is in the research and study phase, has brought the films back to their own family and allowed the Gianese family to regain their own story.

The films of the Gianese archive show on the material of the support and emulsion the effects of the high water, which marked them by its flowing: the images get transformed, sometimes erased, liquified, in other cases corroded. Sections of the films and of the memories contained in them vanish, others miraculously survive intact. Some sections of the films become abstract pictures and suggest new landscapes, with a strong visual impact, real impressions and recordings of the passage of the tide. Today, these films hold within themselves different temporalities: the moment when they were shot, the moments they were hit by the high waters and finally the moment when they were digitized.

If on the one hand the deteriorated images become a creative propeller for artists and film directors, for the family they spur a moment of realization. Seeing these home movies made Giampaolo Gianese, one of the protagonists of the films, reactivate a series of memories about his own experience of the high waters of the Twentieth Century, including the most famous, the Acqua Granda of 1966. But it also brought about a reflection on the vulnerability and resilience of our memories and a parallelism between the dissolution of the memory and the decay of the records.

Venice can be understood as a planetary metaphor (Bevilacqua, 1998). The fragility of the city and its lagoon imprints on this territory the capability to foresee universal issues, like global warming, the rising of sea levels and the life in areas directly affected by these processes. Such fragility becomes, in our opinion, a metaphor of the relationship between home movies and the memory stored in them. In Venice, between the undertow of salty waves and waves of depopulation, this risks vanishing more and more inescapably.

Figure 1: Capovilla family, 1957, 8 mm film. RI-PRESE Archive. © No further use allowed.

Figure 2: Catozzo Capovilla family, 1927, Pathé Baby 9.5 mm film. RI-PRESE Archive. © No further use allowed.

Figure 3: Gianese family, 1950s, 8 mm film. RI-PRESE Archive. © No further use allowed.

Figure 4: Kilmister family, 1978, 8 mm widescreen film. RI-PRESE Archive. © No further use allowed.

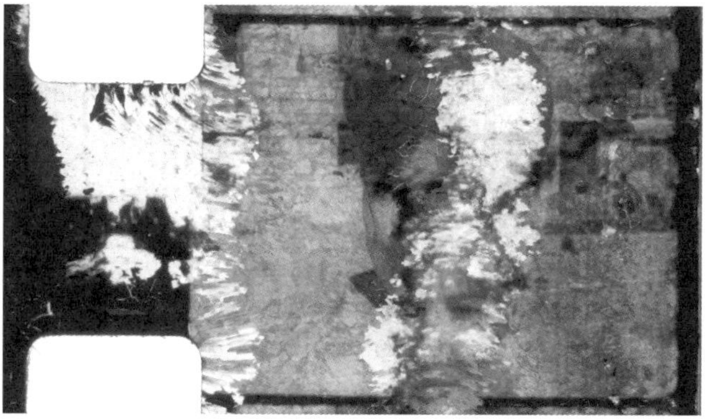

Figure 5: Gianese family, 1960s, 8mm film. RI-PRESE Archive. © No further use allowed.

References

Bevilacqua, P. (1998). *Venezia e le acque. Una metafora planetaria* (Rome: Donzelli Editore).

Del Amo García, A. (2004). Film Preservation. F.I.A.F. Fédération Internationale des Archives du Film Technical Commission. Available: https://www.fiafnet.org/images/tinyUpload/2022/07/English COMPLETO_1.pdf (accessed 10 December 2024).

Morrison, B. (director, 2002). *Decasia.* [film]. Icarus Films.

Morrison, B. (director, 2016). *Dawson City: Frozen Time* [film]. Hypnotic Pictures/Picture Palace Pictures.

Odin, R. (1995). *Le film de famille. Usage privè, usage public* (Paris: Meridiens Klincksieck).

Dijkbouw, Bert Haanstra, 1953

Rommy Albers

Dijkbouw (*Dike Construction*) was the first commissioned film Bert Haanstra made for the Shell Film Unit, the oil company's department that made promotional and informational films. Haanstra was asked in 1951 by Arthur Elton, leader of the film unit, to make a film on the use of asphalt bitumen in the construction of the new dikes in the IJsselmeer.

Dijkbouw gives a clear and straightforward picture of the work of the dike builders who construct the dike body, covering it with woven reed mats and reinforcing it with stone boulders. It's a traditional craft. Then comes a novelty: a large asphalt plate is placed on the dike body and covered with poured asphalt. This further strengthens the dike and at the same time serves as a base for finishing the dike surface.

Haanstra's film fits perfectly in the tradition of 'show and tell'; the whole process is shown in detail and insightfully. Much attention is paid to the worker, from the dragline machinist to the reed mat weaver and porters of stone blocks. The film is also an ode to work, to the anonymous workman. At the end of the film, their boots are drying on wooden poles, like a collection of small monuments.

Dijkbouw is only about the work, about the progress of labour. There is no reference to an outside world. The how and why, the necessity of the work is not shown. It is simply recorded and depicted with great cinematic ability, entirely in the style of the pre-war avant-garde, with great attention to editing, framing and rhythm.

Deltaic Curating of Dialectics

Floris Paalman

Audiovisual media have consistently engaged with spatial issues, affecting spatial practices, policies and plans. Studying audiovisual media as both representations of and agents within spatial developments involves addressing their epistemological and ontological dimensions. This essay focuses on the Dutch delta, examining how audiovisual media contributed to regional development through discourses expressing diverse visions and ideologies. At the same time, it addresses how archives bring such discourses to the fore, introducing the concept of 'collective cognition' to understand their functions and broader implications. Central to this inquiry is the way in which audiovisually mediated dialectics on the environment become manifest in collective cognition. Using and elaborating on the theory of cultural ecology, the article examines factors affecting spatial developments and debates about them. The study focuses on the 1960s-1980s, a period marked by rapid modernization and environmental activism, accompanied by, on the one hand, informational and promotional films about construction projects and industry, and on the other, critical media addressing ecological sacrifice, pollution and nuclear risks. By analyzing archival collections and curatorial practices, the article proposes 'deltaic curating of dialectics' as a framework for understanding and facilitating the role of audiovisual media and archives in shaping spatial and societal dynamics.

Introduction

Audiovisual media have always engaged with spatial issues, from the experience of the city to geopolitics (Mediapolis, 2024). Doing so, audiovisual media have informed people, affecting spatial practices, policies and plans. Studying audiovisual media as representations of spatial developments *and* as agents within them implies both epistemology and ontology. My book *Cinematic Rotterdam* has presented such a view, explaining how audiovisual media have contributed to urban development (Paalman, 2011). I will extend this approach to the perspective of the Dutch delta, where the rivers Meuse and Rhine converge, and diverge again into multiple outlets. Here I will examine how audiovisual media have contributed to regional development, but also address how audiovisual archives are part of this ontology, while affecting what we can know about it.

In film studies, on the one hand, archives usually remain invisible. Studies of audiovisual preservation, on the other hand, mostly concern archival care and management. But Jiří Anger's archival film ontology connects these fields, as 'the aesthetic and figurative function of the moving image is ontologically tied to the material world' (Anger, 2024, 15). Anger demonstrates this through distortions in archival films, which reveal the production apparatus, as well as archives that have prolonged their existence (2024, 98). The apparatus can be extended to the Foucauldian *dispositif*. This broader technical and social-political configuration gives rise to discourse, which is governed by the archive (Foucault, 2002, 145). The archive does so through a memory function, commonly understood as 'collective memory' (Halbwachs, 1925), existing within a specific environment and related to everyday life and politics, as well as 'cultural memory' (Assmann, 1995), which spans many generations and stabilizes society. Both concepts, often used interchangeably, have become paradigmatic in archival and heritage studies (McDowell, 2008; Harris, 2012; Cook, 2013; Apaydin, 2020).

I will rely on Halbwachs' concept for his emphasis on the environment (Halbwachs, 1980, 155). This is opposed to archival scholar Michelle Caswell's view (2014, 45) of decentralized, digital transnational archives that create 'memoryscapes,' a conceptualization inspired by Appadurai's scapes of globalization. Appadurai (2010, 9) argues that 'in thinking about area studies, we need to recognize that histories produce geographies and not vice versa'. But I see a two-way process, along with Dagmar Brunow (2017, 106), who argues that online collections are still locally embedded.

Besides spatiality, another issue is at stake. Memory in the social context is now taken for granted, but in fact it is a metaphor based on the human mind (Errl, 2010, 4; Hedstrom, 2010). Memory in the individual is connected to other cognitive functions, such as evaluation and imagination. I therefore propose to see 'collective cognition' as the principal archival function, building on sociologist Niklas Luhmann (1997); while culture serves a memory function to distinguish between present and past, in order to evaluate and monitor developments, culture also performs the function of 'oscillation', to move across existing categories, creating new ideas and possibilities.

In order to ontologically ground epistemology, I will research environmental discourses, especially dialectics – that is, 'conversations' between oppositional views and voices, 'mediated' through audiovisual media – and how they come to us through archives. This leads me to the question *how audiovisually mediated dialectics on the environment have become manifest in collective cognition*. Mediating here goes beyond relay; audiovisual media also participate in rhetorics. 'Being manifest in collective cognition' concerns archival operations of preservation and access, which bring something into the present again, continuing discourses and implied politics.

The concept of collective cognition aligns with Martina Löw's concept of *Eigenlogik* (2013, 896), which shapes a city's identity, as well as the concept of 'culture core' in cultural ecology, proposed by the American anthropologist Julian Steward. Cultural ecology refers to possibilities set by an environment, which can be actualized through an interplay between ideas and resources (Steward, 1976, 40). The culture core steers cultural growth; I call its regulating instance 'collective cognition.' As a pluralistic mind (Bateson, 1987, 440, 490), it monitors and directs the development of which it is part. Collective cognition is specifically seen here as the implication of archives, including audiovisual media, besides other forms of expression existing within the cultural ecology (Zapf, 2021, 136).

According to Steward (1976, 41-42), cultural growth happens through environmental factors, which I call 'internal factors', and historical influences, such as diffused ideologies or technologies, which I call 'external factors'. Their interplay enables 'sociocultural integration' (Steward, 1976, 43). An expression makes sense at a certain level that integrates particular activities, interests and references. For example, a home movie addresses a household; recordings of community life are meaningful within a town; and television programmes with general cultural relevance are meaningful within a country. Accordingly, there are family archives, city archives and national archives. Collective cognition is manifest at all levels, moderating the interplay between internal and external factors, including cultural and social-economic exchanges, governance and politics. The moderating capacity is ultimately a matter of curating.

Appropriate in the context of the delta is Stefanie Hessler's 'tidalectic curating', based on Brathwaite's 'tidalectics', informed by the tidal action of oceans. Instead of an immediate cause-and-effect relationship, tidalectics follows a holistic view, if always partial and situated (Hessler, 2020, 252). Moreover, Hessler's curatorial approach is 'performatively engaging with the social and ecological structures of the systems in which it operates' (2020, 256). 'Tidalectic curating' proclaims openness, but it is still directional, towards a better understanding of ecological complexity, forging sustainable forms of living. I endorse and articulate this aim (cf. Newell et al., 2017), stressing Laura Marks' (2004) idea of curating being based on an argument, as an ethical responsibility. Yet, I see such rhetorics as a process (Iacob, 2023, 35) and collective endeavour, which draws an epistemological horizon on which to position the case below.

In the Dutch delta, water and nature exist next to industrial areas and cities like Rotterdam, Breda, Middelburg and Vlissingen. This has given rise to a dialectics of sorts. However, the delta has neither administrative status nor institutions of its own. Instead, there are different provincial and water authorities (Unie van Waterschappen, 2024). This compromises representation and delineation. Alternatively, one could consider the Netherlands as a whole. Dutch culture lives on water, as many Dutch film classics exemplify, from the first Dutch fiction film, *De mésaventure van een Fransch heertje zonder pantalon aan het strand te Zandvoort* (Mullens, 1905), to the social drama *Dead Water* (Rutten, 1934), the classic documentary *The Voice of the Water* (Haanstra, 1966), and the epic feature *Admiral* (Reiné, 2015), to name a few. They serve the cultural memory of the nation.

Within the Delta region, Zeeland is the main province. In its capital, Middelburg, both the Zeeuws Filmarchief and Film Festival Zeeland were established in 1989. Together they called for historical films (*Provinciale Zeeuwse Courant*, 1990). While acquiring and showing them, an inventory was made, published in 1992 (Jongejan, 1991; Zeeuws Filmarchief, 1992). The festival continued and archival presentations remained part of its programming (*NRC Handelsblad*, 1992). Eventually it became the international festival Film by the Sea, in 1999 (*De Bevelander*, 1999). The film archive, in its turn, did not become an institution. Collected films went to the Zeeuws Archief, and in 1997, Omroep Zeeland started the archival compilation series *Trugkieke*, directed by Jan-Willem Antheunisse, one of the initiators of the film archive (Wij zijn de stad, 2018). About 350 of its episodes have been published on YouTube, from local amateur recordings to footage from national collections. The latter have mainly informed this study, especially collections at the Netherlands Institute for Sound and Vision (Beeld&Geluid, 2025) and Eye Filmmuseum (Eye, 2025), and the aggregate database Media Suite (Media Suite, 2025), which offers selections from Dutch institutions online, accessible to students and researchers. A few other collections, platforms and YouTube channels have been considered too, including EUscreen (EUscreen, 2025), an open-access database with audiovisual collections from all member states of the European Union. For contextualization, I have used the newspaper databases Krantenbank Zeeland (2025), Tijdschriftenbank Zeeland (2025), and Delpher (2025).

This research focuses on the 1960s-1980s, as a period marked by accelerated modernization and activism, in the light of environmental issues. On the one hand, I consider in this paper informational and promotional films about industrial developments and construction projects. On the other, I consider critical media, addressing problems such as nature being sacrificed for industry, pollution and the risks of nuclear energy. I will present their dialectics, and in conclusion I will propose the concept of 'deltaic curating of dialectics'.

From Safety to Spectacle

After the flood of 1953, the Delta Plan was developed. Many state-sponsored films recorded the execution of the Delta Works in the period 1953-1986 (such as the films made by Hoving, 1958-1961; Batchelor, 1961; Haanstra, 1962; Breijer, 1969, 1974; Fernhout, 1968; Burcksen & Herblot, 1985-1988). All of them call attention to the ingenuity and power of the works, for which safety is the main argument. This is already clear from Haanstra's *Delta Phase 1*, about the closing of the Veerse Gat. The film shows, as its narrator says, 'the first small link in a chain of grand works that should connect many islands into a close-knit and safe delta land' (translated by the author). The catalogue of Sound and Vision lists several entries to *Delta Phase 1*, without cross-referencing. One of them says: 'The film is introduced with images that suggestively show Zeeland as a land where life is dominated by the sea with all its evil sides. There are some poignant shots of the 1953 flood' (Beeld&Geluid 2025a; translated by the author). This description reproduces the rhetoric of sea-as-enemy. The film is also preserved by Eye, whose catalogue says that 'protests from the Veere fishermen . . . are drowned in the euphoric images of the festivities on the occasion of the closing of the flow hole. The fact that the festivities are somewhat delayed by the sudden fog only increases the build-up to the film's dramatic climax – and is certainly not processed as a critical comment' (Eye, 2025a; translated by the author). These descriptions dialectically position the film.

While not publicly available through Sound and Vision and Eye, *Delta Phase 1* has been put on YouTube, in low quality, by a user named mdftrasher (2023). A viewer asks if the film could be uploaded in HD; mdftrasher says 'no', as it comes from a VHS recording of a television broadcast in 1996, part of the retrospective, *De films van Bert Haanstra* (Haanstra & Van Fucht, 1996). A video glitch at the bottom and the channel's logo at the top reveal this provenance, creating 'weird shapes' (Anger, 2024, 2). There is another remnant: a part of an interview with Haanstra, before the film. In it, Haanstra speaks about a shot of only four seconds. It sits at the film's climax, when the dam is being closed, witnessed by spectators and (foreign) cameramen, as the film shows – which is a self-reflexive image of sorts. When the job is done, the man in charge 'takes his cheap camera to make a snapshot of the [closed dam], for his album, for mom', as Haanstra says with mild irony. Within the broader *dispositif*, the 'prefix' with Haanstra speaking becomes a 'weird' self-referential artefact; the act of taking a snapshot, recorded on film, broadcast on television, taped on video, digitized and shown on YouTube, overlaid with commercials, shows a layered ontology that corresponds to Brunow's understanding of cultural memory being 'constantly reworked through processes of remediation' (2017, 98). The gesture of the snapshot documenting a reality is remediated and multiplied; eventually, it makes Haanstra's irony applicable to his own film as well.

The end of *Delta Phase 1* shows a new park with holiday homes. The narrator says: 'the desire for quietness and freedom of the city dwellers from one of the most densely populated parts of Europe will take off in this Old Zeeland' (translated by the author). Due to the dams and bridges, 'Old Zeeland' becomes accessible for city dwellers, which is also celebrated by *Polygoon Journaal* (1969-32) through a remarkable time-lapse motion. Its impact is the subject of the documentary *Het proces Renesse* (Schaper, 1966), directed by Rotterdam-based filmmaker Jan Schaper and producer Jan van Hillo. From a realist perspective, they observe how Zeeland changes: youths from Rotterdam flock together on beaches, spending the night in clubs. The narrator says: 'The island makes the youth overconfident. The wind and the sun, the salty smell of the water. All this creates excitement in them. It makes them feel happy and tempts them to do silly things, so that many girls between 15 and 18 grow up here in one summer' (translated by the author). This reality was unacceptable and moralists at the Christian broadcaster NCRV banned the film, as explained by the film's sound technician and producer Christine van Rhoon (conversation with the author, 2010).

At the same time, there was criticism regarding the environmental impact of the Delta Plan (Zeeuwse Ankers, 2024). Fishermen, activists and academics joined forces in 1968. Their view gained momentum in the mid-1970s, through news reports, documentaries and educational programmes like the *Bedreigde landschappen: De delta* (Teleac, 1976). With the advent of a progressive government led by Joop den Uyl in 1973, the planned dam was changed into a *pijlerdam* ('pillar-dam'), called Eastern Scheldt Dam (Oosterscheldekering). This allowed for tidal action, hence preserving the unique natural area. However, current affairs programme *Aktua* (TROS, 5 October 1982) emphasized the budget gap of 700 million guilders, with interviewees blaming the government for mismanagement.

In 1986, the dam was inaugurated by Queen Beatrix, broadcast live on television by NOS for four hours. It is one of the few programmes about the Dutch delta available on EUscreen. One reporter recalls the discussion on the expenses, but explains that there was no way back. In the end, the dam was heralded as the jewel of the Delta Works, notably by TROS, broadcasting the film *Delta Finale* (Burcksen & Herbot, 1988). The film's motto is: 'A barrier for safety and environment,' which echoes *Delta Phase 1*, while adding 'environment'. The title was affected by dialectics, spinning the heroic story of the Delta Plan which attracted ever more interest. During the construction period (1979-1986), an information centre was therefore established by Rijkswaterstaat on the work island Neeltje Jans. It showed films, especially *Delta Finale*, in three successive versions (*Dutch Delta*, 1983; *Delta Finale*, 1985; *Delta Finale*, 1988).

In Sound and Vision's catalogue, the final version of *Delta Finale* is listed four times, under different titles, without cross-referencing. The reason is that film prints have come from different collections, but this is unclear for users, which obscures their provenance and positionality. One entry (with 109 analogue carriers) mentions that the film was commissioned by the ministry of Transportation and Water Management, with the genre indicated as 'informational film' (Beeld&Geluid, 2025b). The other entries barely have any metadata, but they are indicated as 'company film' (Beeld&Geluid, 2025c, 2025d, 2025e). However, one of them (2025d), still mentions broadcaster TROS, suggesting that the film comes from the television collection. Surprisingly, under the series title 'Bedrijfsfilms Baggerbedrijf Van Oord' (company films for dredging company Van Oord), a digital copy is made available in Media Suite, left unmentioned elsewhere. With these confusing (series) titles and no integration of records, access is compromised.

However, *Delta Finale* is also available on YouTube (Kerker, 2015). There are 79 comments, some concerning the film as such (translated by the author):

> @gwvandersnel [2 years ago]
> Thanks for posting this film! During the construction of the Eastern Scheldt Dam, this film was shown to visitors of the construction of the Eastern Scheldt Dam in a temporary building on the work island. I loved the movie then and I still do now. I was involved in the construction for a year and a

half, so on the work island itself, our company took care of the control of the gates. It was a fantastic time together.

> @jansmalheer4514 [7 years ago]
> Film was shown for years in the temporary information centre at Neeltje Jans. I worked there for 4 years and the film and music bring back a lot of memories! A wonderful time with a very special job!

These quotes attest to a 'double memory': of what is being shown and of the film's exhibition, as part of the spectacle of the Delta Works. In some comments, the subject and the film become vehicles to convey life stories:

> @bernardkannekens7286 [1 year ago]
> My father worked on the construction of the storm surge barrier in the Eastern Scheldt from start to finish. On the one hand, it is very symbolic, because my mother had to flee from the flood at the age of 14. The dike broke where her parental home was and they lost everything. On the other hand, we as a family have fond memories of the construction of the barrier. We regularly went to look and saw the barrier grow. In the weekend, we even sometimes went to 'stay' on the work island Neeltje Jans. We slept in a construction shed and had the island and beach to ourselves for a weekend. You never forget that. Dad and mom are no longer alive. Of course, we had this film on a video tape at home. It has been a long time since I have seen it. It brings back many memories. We are still so proud of the construction of this barrier!

The writer explicitly mentions the video tape of the film, which is another instance of 'double memory'. Besides sharing memories, some people comment on the film itself, praising its technical depth, while most respond to the film's content, for example:

> @antoinjanssen6875 [11 months ago]
> It's still a pity that we didn't use that strong current for a tidal power station.

This remark concerns an unfulfilled potential, which is both retrospective and prospective; memory enables oscillation. A tidal power station is a historical phenomenon, like the sixteenth-century tidal mill in Middelburg (Molendatabase, 2023). Hence, history informs future prospects.

The information centre became a tourist destination, and developed into the theme park Deltapark Neeltje Jans, which is still showing these kinds of films. When *Delta Phase 1* stated that through the Delta Works new land was being created for industry, buildings and leisure, Haanstra may not have anticipated that this land would eventually accommodate screenings of films like his own – an ontological closed circuit. The shift from the rhetoric of safety through engineering to the presentation of the Delta Works as spectacle is a historical act of oscillation.

Future Horizon

Browsing Eye's catalogue, a striking instance of oscillation appears: Bob Visser's experimental science-fiction film *Plan Delta* (Visser, 1989), or so it seems, as the film itself is inaccessible. Its manifestation can only be reconstructed through written reports. The film tells a remarkable story, but its own story is similarly remarkable. Shot on Neeltje Jans, it was screened there too, in control centre Topshuis, which anchors the film ontologically. It was a special event of the first edition of Film Festival Zeeland in 1989, meant to stimulate tourism (*Algemeen Dagblad*, 1989). For the screening of *Plan Delta*, a shuttle bus brought spectators from Middelburg to Neeltje Jans. A second screening took place in Electrobioscoop-Schuttershoftheater, Middelburg (Van Damme, 1989). The film literally moved between Zeeland's old centre of trade and ship-building and Neeltje Jans, epicentre of the region's renewed identity of deltaic engineering. Oscillating between past and future, *Plan Delta* tells yet another story, through politics of the imaginary.

The film project began as a documentary production in 1983. Bob Visser, based in Rotterdam, heard about the planned inundation of the *pijlerdam* and rushed there to make recordings. He edited them into a seven-minute short for TAPE-TV (VPRO, 1 February 1984). Being born in Zeeland, Visser thought about the claim that the dam would protect the land for the next 250 years (Veld, 1989, 20). More recordings were made on Neeltje Jans, supported by Rijkswaterstaat, and in a studio in Rotterdam, Utopia Hal 4. In 1986, a draft was shown at the Nederlandse Filmdagen festival in Utrecht. *De Volkskrant* reported on it, notably in its section 'Science & Society'. Visser is quoted saying that the Delta Plan has been presented to the Dutch as the summum of engineering. The result is that people are no longer afraid of the water and think they can conquer everything (*De Volkskrant*, 1986).

Later in 1986, the film was made into a theatrical show, with live music by the Californian new wave band Tuxedomoon. Band member Bruce Geduldig acted on stage as the film's narrator. The show was performed six times, in different theatres, including Rem Koolhaas' Danstheater in The Hague (still under construction at the time), where Visser filmed the projected images and performance. The recordings were used again in the film, as temporal layers for time travel (Veld, 1989, 20). The show received critical acclaim; Visser got additional funding (in total 350,000 guilders), and the final film eventually premiered at the International Film Festival Rotterdam in 1989 (Slavenburg, 1989).

The story goes as follows: anno 2330, the lowlands are flooded. People live in an elevated bunker city, which is the *pijlerdam*, divided into different communities, each living in its own pillar, in apathy. They have gradually lost their virility, and need the help of two children, who were saved by an angel during the flood of 1953. In order to travel back and forth in time, people sit in a theatre and watch archival footage of the flood. An expedition is organized to search for healthy people, living in primitive conditions on the island Neeltje Jans. In the meantime, the bunker city is threatened by water. An alien fleet comes to safeguard the city, and the fleet's commander, who is also the film's narrator (played by Bruce Geduldig), calls out to the people, but they don't cooperate (Slavenburg, 1989).

Visser drew inspiration from British author Doris Lessing's science-fiction novels, the series *Canopus in Argos: Archives*, published 1979-1983 (Jager, 1989). He appropriated this 'external influence' for a Dutch context, where he saw the struggle against water as historically decisive (Veld, 1989, 20). This exemplifies how 'Science fiction displays a dialectics between reality and (im)possible experiences of the world . . .' (Monticelli, 2019, 5). Visser also dwelled on Lessing's religious tropes, which were relatable to him through the legacy of his father – who also sponsored his film – including books about predestination and the conversion of people in Zeeland, published around 1900 (Jager, 1989). Visser was intrigued by the books' archaic, moral language, which he projected onto future speech. Calling his film 'time fiction', history informed a future imaginary (Figure 1).

Journalist Peter Slavenburg appreciated the film as much as its theatrical iterations, which he called 'magnificent, exceptionally impressive performances'. He added: 'although less expensive than the Delta Plan, the film is almost equally grandiose' (1989; translated by the author). But responses varied. Film critic Peter van Bueren (1989) called it an 'incomparable and remarkable film', and commented on its stream of images, writing: 'Stand on the beach in a storm and let a giant wave wash over you. This experience is somewhat like undergoing *Plan Delta*' (1989; translated by the author). Several other critics could not make sense of it (Kotterer, 1989; Kok, 1989). Slavenburg stressed that the film is actually about safety, and he quoted Visser saying that:

> We can protect ourselves against the water, but the question is: for how long? In 1953, the idea of environmental pollution hardly existed yet. We built dams, but at the same time peo-

> ple messed around. Now, less than forty years later, we have a hole in the ozone layer and we are threatened by the greenhouse effect. So it will never stop: the water will always rise again. It is the consequence of our way of life, which forces us to leave, sooner or later, exactly like the people in Bangladesh will have to leave the delta. In that sense, the Delta Plan is no different than the sandbags used in 1953 (Visser in Slavenburg, 1989; translated by the author)

In the 1980s, few people imagined that today water levels would rise so quickly that by the twenty-third century the Netherlands will be flooded indeed (Bregman, 2023). And few people saw this as a global problem the way Visser did. His reference to Bangladesh is telling; it was informed by Dutch television frequently reporting on floods there, in 1984, when Visser developed his idea (e.g, EBU, 28 August 1984; NCRV, 16 November 1984; NCRV, 21 December 1984), and just before the film's release (e.g., TROS, 6 September 1988; NOS, 11 September 1988; NOS: 13 September 1988; VOO, 21 September 1988). This is another 'historical influence'. *Plan Delta* is the result of memory and oscillation, by moving across times and places, categories and ideas, drawing a future horizon.

Visser extrapolated the issue of safety to the extent that another way of thinking is required. Only now there are broad campaigns advocating change. WWF writes: 'The delta works protect us against flooding, but they have also disrupted nature in the delta' (WWF, 2024; translated by the author). Through collective cognition, a new paradigm must be developed.

The Shock Effect

Along with modernization and industrialization, pollution had become part of the discourse, fuelled by the documentary *We stinken er in . . .!* (1970) – the title meaning both 'we get fooled' and 'this stinks'. It was made by director Jan Van Hillo, born in Maassluis, in the delta region, in collaboration with researcher Wolf Kielich and cinematographer Jan Schaper, with whom Van Hillo also made *Het proces Renesse* (Schaper, 1966). *We stinken er in . . .!* was produced by Christian broadcaster NCRV, which believed that God had entrusted humans to take care of the Earth. The documentary starts with the 1953 flood, saying that there had been critical voices before 1953. Other disasters are similarly awaiting us if nothing will be done. Van Hillo conceived of the documentary when directing the documentary *Mijlpaal in het Haringvliet* (1970), about another 'milestone' of the Delta Plan. In turn, *We stinken er in . . .!* contends that humans have disturbed nature's balance. The film alternates between developments in the Netherlands (e.g. Rotterdam) and abroad (USA and Japan), and between threats and perspectives, while interviewing experts, such as Barry Commoner.

The documentary had an impact, but television was ephemeral. The television 'archive' back then was merely a repository and not accessible. It made Van Hillo and Kielich decide to publish the film script as a book, with explanations about their decisions, additional information, quotes and biblical references. The book (Van Hillo & Kielich, 1971a) became an annotated stand-in for the film, and an archive of sorts (Figure 2). Due to the shock effect of the film, a sequel was made, *Het uur van de waarheid (The Hour of Truth*; Van Hillo, 1971), which was immediately accompanied by a book (Van Hillo & Kielich, 1971b). *We stinken er in . . .!* was eventually developed into a series of five documentaries (Van Hillo, 1970-1971), with many prominent speakers, such as Alvin Toffler, in an episode based on his book *The Future Shock* (Van Hillo, 1970). The series became a landmark in NCRV's history (NCRV, 1 September 1989; NCRV, 2 December1994). The series is available in Media Suite, but there are inconsistencies and omissions in the titles, with limited metadata, and the books are not mentioned. This stands in contrast to the series' impact.

While showing developments across the world, *We stinken er in . . .!* was ontologically embedded in the delta region, where Van Hillo conceived the series, and to which he linked back. The second episode, for example, shows pollution in nature reservation area Biesbosch, which was highlighted in a local review (*De Stem*, 1971). In this way, the series directly informed discourse in the region, propelled by newly established organizations, and existing ones, such as the Bond van Plattelandsvrouwen (Union of Rural Women). In November 1970, this union organized a study day in Goes, with expert talks about 'environmental hygiene' and the responsibility of housewives (*De Stem*, 1970), together with an exhibition. It also showed the film *Om het laatste groen* (IVN, 1965), distinguishing welfare from well-being (IVN, 2024b). The documentary series facilitated discussions on industry releasing toxic gases into the air and waste being discharged into open water by local enterprises and towns, as reported earlier that year (AVRO, 14 July 1970).

The issue of pollution in the delta was further elaborated by director André Truyman, also collaborating with Jan Schaper, in the documentary trilogy *Het land van de dwazen* (*The Country of Fools*), presented by catholic broadcaster KRO. The second episode (André Truyman, 19 January 1971) starts with speeches by a veterinarian and a politician, on the impact of industrial pollution. The narrator continues: 'What is happening here looks like a new kind of colonization. After all, the population itself does not seem to be directly involved in the decisions taken by the government regarding industrial establishments' (translated by the author). But he also explains that most Zeelanders 'suffer from paralyzing fatalism and don't resist'. These fragments have been included in an episode by *Trugkieke: Zeeland in de Hilversumse kijker VI* (Omroep Zeeland, 25 May 2020), on Zeeland's presence on Dutch television throughout the last 50 years. Broadcast in 2001, and digitized and made available on YouTube in 2020, supported by the Province of Zeeland, it shows archival activation, remediation and appropriation of the discourse within the region.

In 1974, environmental organization Vereniging Milieuhygiëne Zeeland began publishing a monthly magazine, *De Gouden Delta* (*The Golden Delta*), with texts, photographs and cartoons, criticizing technocrats and industrialists. It spearheaded the environmental discourse in Zeeland (De Jong, 1984). Connections were drawn between industrialization, consumption culture and pollution, an argument already made in the television documentary *Polders voor Industry* (*Polders for Industry*; Van der Velde & Smits, 1961), which was critical of the extension of the port of Rotterdam (Paalman, 2011, 426). Political parties also joined the discourse. The film *Socialistiese Partij* (1976) explicitly addresses both environmental and social problems in Rotterdam and elsewhere. It is preserved by the Rotterdam City Archive, where it is available for consultation on-site, but it has become detached from related print media and archival records. As it is often the case with audiovisual media, this compromises the archival principle of *respect des fonds*, 'in which all records originating from an administrative authority, corporation, or family would be brought together in a fonds based on their origins and function' (Jimerson, 2009, 72).

Such archival disconnection is also exemplified by the 'death' of the magazine *De Gouden Delta,* after ten years of struggle. Calling attention to their cause once more, the editors staged a 'funeral' on All Souls' Day, 2 November 1983. NOS television came to record the 'cremation' of the magazine, which was put in the water of the Eastern Scheldt, floating away while burning (NOS, 9 November1983). This television report is now available online in the Media Suite, but the magazine itself is not easily retrieved from the hereafter; it is only available on-site at the Zeeuws Archief (Zeeuws Archief, 2002). As a result, the television report exists in isolation, having in fact eclipsed the magazine and its history. This archival dimension could not have been anticipated by the NOS television reporter when he asked if the death of *De Gouden Delta* meant the end of the environmentalist movement. An editor of the magazine responded that a new movement would emerge to counter the growing problems in which interrelated factors were at play. And indeed, in 1984, de Zeeuwse Milieufederatie (Zeeland Environmental Federation) was established, in which various previous organizations got united (Zeeuws Archief, 2002).

Television programmes would continue to address environmental issues. Artists also contributed to the discourse, like the musicians Leo Cuypers and the Willem Breuker Kollektief in the film *Zeeland Suite* (Hulscher, 10 January 1978), addressing the uniqueness of the land. Alternatively, activist film collections should potentially be considered too, such as those of Greenpeace, which are stored at the International Institute of Social History in Amsterdam (IISG, 2024). However, this collection still awaits proper archiving, like many activist media archives elsewhere, due to their non-conventional nature.

Changing Fights

With industrialization and urbanization, more energy was needed. Nuclear power offered it. In various countries, including the Netherlands, films were made to explain the subject, such as *Atomos* (Erends, 1960), a promotional film for Philips, with puppet animations of Galileo Galilei and Einstein, in which Philips devices were used for the generation of nuclear power. However, after the initial enthusiasm, safety became an issue. Films were subsequently made about responsible storage of nuclear waste, like *Project Salt Vault* (Atomic Energy Commission, 1969), one of the many films commissioned by the US Atomic Energy Commission. Both films are part of Eye's collection, although inaccessible – yet, many such films are made available online elsewhere (Touran, 2007-2025).

While the Delta Plan was built on the rhetoric of safety, authorities refrained from such language when building nuclear reactors, including the one in Borssele, Zeeland (1969-1973). It was meant to deliver cheap energy for the French aluminium producer Péchiney, to accommodate its establishment in Vlissingen (*Algemeen Handelsblad*, 1968). Being part of a larger industrial development scheme, it raised criticism for destroying the landscape and for environmental pollution. Dutch television covered such issues over an extended period of time. For the period 1960-1989, Media Suite yields 123 results for queries concerning Zeeland and environmental topics (terms searched: 'milieu', 'waterverontreiniging', 'luchtverontreiniging', 'verontreiniging', 'milieuverontreining', 'vervuiling', 'milieuhygiëne', and 'milieuvervuiling'). Tropes in the discourse were the lack of transparency in how decisions were made, the lack of citizen participation, secret agreements and risks not being communicated.

Before the reactor's opening, television reported on activists handing out flyers to residents to make them aware of radiation (KRO, 12 May 1973; NCRV, 15 May 1973). After the opening, television programme *Brandpunt* (KRO, 13 January 1974) paid attention to the disaster plan. The plan was explained dialectically, through interviews with Borssele's technocratic mayor and an environmentalist, but the report seconds the latter through audiovisual rhetoric. For example, when the plan is explained, a desolate cemetery is being shown and an alarm is heard. Travelling shots show empty streets, as if residents are already hiding in their homes. Moreover, the images are recorded with backlight and solar flares, having an alienating effect. The disaster plan was also ridiculed by *Farce Majeure* (NCRV, 26 January 1974). These three programmes make for a well-curated triptych, presented by *Trugkieke: Kerncentrale Borssele, 1973/74* (Omroep Zeeland, 2020). However, *Trugkieke*'s episodes barely garner comments by viewers; the curatorial potential of propelling discourses is not yet exploited.

Over the course of the 1970s, worries increased; the reactor in Borssele frequently featured in television reports. Different views were presented (e.g. *Op uw gezondheid: Kernenergie en stralingsgevaar*; TROS, 23 January 1974), but scepticism grew, especially with the construction of a breeder reactor in Kalkar, Germany, close to the Dutch border. The Dutch Anti-Kalkar Komitee organized a mass demonstration in 1977, which attracted media attention (besides television reports there is also for example the film *Demonstratie Kalkar, 1977* by Heerkens, 1977).

Worries culminated when in Harrisburg, Pennsylvania, a partial nuclear meltdown took place on 28 March 1979, the so-called Three Miles Island Accident. Radioactive gases and radioactive iodine were released into the environment, which caused anxiety around the world. For multiple days, Dutch television reported on it (e.g., NOS, 1 and 2 April 1979; VARA, 3 April 1979; AVRO, 5 April 1979), sometimes with alienating imagery and eerie electronic soundscapes (e.g., TROS, 7 April 1979). The Anti-Kalkar Komitee, renamed Landelijk Energie Komitee (National Energy Committee), pointed to similarities between the plants in Harrisburg and Borssele and announced a demonstration, demanding the closure of Borssele's reactor. Socialist broadcaster VARA responded to their call (*VARA-Visie*, VARA, 3 April 1979), thus helping to mobilize people.

The demonstration in Borssele was reported by the daily news report (NOS, 7 April 1979). It is available in Media Suite, but without comments, which were only read live during broadcasting. The remaining 16 mm footage has become a typical archival television artefact, which resonates with Jiří Anger's archival film ontology, for its indexical relation with the apparatus. Media Suite as interface remediates this artefact without mentioning its status. Moreover, an 'archival reversal' is being created here, as cinema newsreel *Polygoon Journaal* (Polygoon, 1979-16) has instead been preserved and catalogued properly, while by 1979, cinema newsreels had already been less important than television news for two decades. Polygoon shows the demonstration, with comments by participants, followed by background information on the plant's safety measures, which implies a discursive position. Sound and Vision has posted it on YouTube, with 23 comments that lack any moderation. However, it has also become embedded in the reactor's Wikipedia page (Wikipedia, 2023), which sketches a background.

The demonstration became a critical reference in the ongoing discourse (see e.g. *Villa VPRO*; VPRO, 20 February 1980). Foreign films contributed to this as well (De Ruiter, 1976), especially those distributed by Cineclub Vrijheidsfilms, such as the American production *Lovejoy's Nuclear War* (Keller, 1975), and the German production *Lieber heute aktiv als morgen radioaktiv* (*Better Active Today Than Radioactive Tomorrow*; Gladitz, 1976). These militant films are preserved by the International Institute of Social History in Amsterdam. Although its catalogue does not link films with paper records, the materials still enable a reconstruction of the discourse.

The main target of the anti-nuclear movement became the reactor in Dodewaard, situated along the river Waal, feeding the delta. On 19 October 1980, 15,000 people came together for a blockade. Besides television reports, a one-hour-long film was made about it, *Dodewaard gaat dicht: Blokkade van de kerncentrale Dodewaard* (*Dodewaard Is Closing: Blockade of the Nuclear Reactor Dodewaard*; Jansen & Elzinga, 1981), which is part of the Cineclub collection. Shot on 8 mm and distributed on 16 mm, the film points to an alternative media practice, at the margins of a 'parallel cinema' (Croombs, 2019, 71-72). This ontology attests to the film's current status too – unavailable, without metadata. However, an article in the communist newspaper *De Waarheid* (1981) makes clear that this film served a training purpose for the next blockade. That happened on Saturday 19 September 1981, when 15,000 people showed up again. The police carried out charges, but the blockade lasted for three days. The next Saturday, 40,000 people gathered to protest against both nuclear power and police violence (Laka, 2024).

Two other films deserve mention here. The first is a mid-length feature, *Fiction* (Holthuis, 1981), which premiered at the International Film Festival Rotterdam. It is about the 'fading borders between reality and fiction, caused by the anxiety and uncertainty concerning nuclear energy', as described in Eye's catalogue, where the film is available on-site. The story is about a man working at a nuclear power station who experiences an accident due to managerial negligence. He then instigates a serious accident to provoke responses. A publicity leaflet for the film says that there is too much information about nuclear energy, which complicates an objective conclusion, but the main problem is that the 'atomic lobby' is in charge. It means that radioactivity will stay and that one must find a way to live with it. Rather than an activist intervention, the film is an artistic reflection upon the discourse and the anxiety fuelling it.

The second film is *Armoede van de toekomst* (*Poverty of the Future*; Entrop, Hopmans & Siegmund, 1981), an educational production to enable discussions among young people about nuclear energy (Figure 3). It was developed by film collective Solidariteitsfilms from Breda, with support from activist fund X-Y. The film starts with the demonstration against the reactor in Borssele in 1979 and then shows high school pupils discussing energy resources. In an interview with a regional newspaper, Marcel Siegmund explains that the film was made for and by young people, and that Solidariteitsfilms merely enabled them to make this 'youth newsreel', lasting almost an hour. He adds that: 'This has clearly become a newsreel against the use of nuclear power. The youth fully agreed that enough money is already spent on propaganda in favour of nuclear power' (*Provinciale Zeeuwsche Courant*, 1981, 9; translated by the author). *Armoede van de toekomst* is the second title in a series of 'youth newsreels', made after *Erbij horen* (*Belonging*; Hopmans, 1980), about consumption and peer pressure, shot in Zeeland too. The newsreels were shown in more than 250 places, at schools and youth centres, in Zeeland and elsewhere (*Provinciale Zeeuwsche Courant*, 1981, 9).

Solidariteitsfilms also made *Speelgoed voor later* (*Toys for Later*; 1981), a fiction short about toys being ideologically charged, casting kids in gender roles and prescribing social behaviour. Highlighting things playfully, the film shows a toy manufacturer who wants to keep up with the times. He produces a nuclear reactor model, together with miniature protesters and a police squadron. Through film, youth learned how they had unconsciously learned things as kids. In this way, Solidariteitsfilms raised awareness about interconnections between social and environmental concerns, and between industrialization and consumption. This fostered critical thinking and political engagement, which affected actual developments. While the reactor in Borssele is still operational today, Kalkar's reactor would never become operational and Dodewaard's reactor eventually closed down in 1997 (Laka, 2024).

Extended Cultural Ecology

Interconnections between social and environmental concerns are not confined to one place. This causes a challenge for archival curating, which becomes clear when an extended cultural ecology is considered. As Visser (1989) referred to Bangladesh when discussing his film *Plan Delta*, one can examine, for example, how the two delta regions relate.

The Zeeuws Archief (2024) shows how Zeeland, and especially its capital Middelburg, took a prominent position in colonialism. During the existence of the Dutch East India Company (VOC), 1602-1799, its wharf in Middelburg built 336 ships that would make a total of 1147 voyages to the east. This included Bengal, where the VOC established the Dutch Directorate (Prakash, 1985, 40). It was an important source for fine cotton and raw silk, much of which was shipped to Middelburg (Wegener Sleeswijk, 2024, 40). Dutch Bengal continued to exist until 1825, when the British took over, according to the Anglo-Dutch Treaty of 1824 (Bhaduri, 2020, 60). Still, the Dutch stayed in touch with the British; when slavery was abolished in the Dutch colony of Suriname in 1873, indentured labourers were recruited in British India, including Bengal, until 1916 (Gowricharn, 2024, 20). About half a century later, the Dutch returned once more, and Zeeland played a prominent role again.

Having gained experience with the Delta Plan, Dutch engineers started working in Bangladesh (Thomassen, 1975). What began as developmental aid in 1972, soon developed into a business that 'is good for our economy' (Eversdijk, in Paumen & De Ruiter, 1984). International charity campaigns raised millions of dollars for projects in Bangladesh, for which Dutch companies competed with Korean and Japanese companies (Van der Peijl, 1985; De Ruiter, 1985). The Dutch state also played an active role in getting orders for Dutch companies. In this period, Dutch television broadcast myriad programmes about events and developments in Bangladesh, including politics, floods and construction projects. For the period 1960s-1980s, the query 'Bangladesh' in Media Suite delivers 371 audiovisual records – and ten times as much if later periods are included too (3,677 audiovisual records and another 6,345 audio records). However, no context or framework is provided.

Strikingly, on EUscreen, with audiovisual collections from the entire EU, the query 'Bangladesh' yields only seven programmes. One is from Sound and Vision, *Nederland herdenkt en zingt* (*The Netherlands commemorates and sings*; Fijnvandraat, 2003), produced by evangelical broadcaster EO, to commemorate the 1953 flood. In this programme, framed around a memorial concert in the St. Jacobskerk in Vlissingen, extensive archival footage is included. At the end, an eight-minute report is shown about Bangladesh, showing a flood in 2002, with the purpose to raise funds to build shelters in Bangladesh. This is put in perspective through footage from earlier floods there, drawn from Dutch television. This becomes an archive of sorts, but the Bengali footage lacks metadata. Most remarkable, the English summary of the programme does not even mention Bangladesh. Furthermore, I have been the first viewer to see this programme (and many others) on EUscreen, which shows that EUscreen struggles to meet one of its key aims, to serve the 'cultural memory of Europe' (EUscreen, 2025).

Other curatorial projects are needed to understand extended cultural ecologies, but also to provide counter-perspectives. This could happen through the international film festival circuit, where no integrated database exists yet. Searching for Bangladesh on the website of the International Film Festival Rotterdam (IFFR, 2024) yields 20 films, such as *Dukhai* (Islam, 1997), which deals with the Bengali delta and was made with support by IFFR. While the 'archive' of the International Documentary Festival Amsterdam (IDFA, 2024) is harder to navigate, one can still find two films by Kamar Ahmad Simon, *Are You Listening!* (Simon, 2012) and *Day After...* (Simon, 2021), both dealing with the Bengali delta and produced with IDFA's support. They present different epistemologies, showing how people live *with* the water. These films call for another dialectics and another curatorial approach, like Hessler's tidalectic curating, following the oceans that connect the deltas within the extended cultural ecology.

Conclusion

In order to answer the question how audiovisually mediated dialectics on the environment have become manifest in collective cognition, I have elaborated on Halbwachs' collective memory and Steward's cultural ecology, examining audiovisual archival material concerning the Dutch delta. Although archives hardly provide discursive positions of records, they have still allowed dialectics to become manifest. They are multifold, between technocrats and environmentalists, realists and moralists, artists and industrialists. At the same time, there have been dialectics between moving images and the printed press, but also between audiovisual media, such as promotional films versus television reports. Promotional films on the delta did not directly respond to critics; they set their own agenda. They heralded ingenuity and created an image of wonder. But critical perspectives still became manifest. Sometimes this caused change, as with the Eastern Scheldt Dam, but in other cases, like the Borssele reactor, managers adjusted their tactics.

Dialectics are enabled by archives (Foucault, 2002), while 'becoming archival' in turn. They rely on and propel functions of memory and oscillation (Luhmann, 1997). As such, dialectics are both directional and cyclical, through remediation (Brunow, 2017), as exemplified by (the YouTube versions of) *Delta Phase 1* and *Delta Finale*, and through combining tenses. The latter is exemplified by *We stinken er in . . .!* and *Plan Delta*, which projected historical references onto the future, but rather than rejecting the prevailing rhetoric of safety, they shifted the focus to consequences of our way of living. This is a matter of actualization, which happened at different levels (city, region, country, world), but also between them. Industry was no longer confined to Rotterdam, but spread across the region, which caused tensions. Similarly, urban youngsters came to Zeeland for leisure, which caused tensions too (*Het proces Renesse*). Global and local developments have similarly interacted (*We stinken er in . . .!*).

In cultural ecology, the culture core steers internal factors and channels external ones, and moderates their encounter. The resulting integration already happens in individual films, like in *Plan Delta*, in which archival footage serves as future memory, combining internal factors with ideas from Lessing's *Canopus in Argos: Archives*. Integration eventually also applies to entire archives. Both Eye Filmmuseum and the Netherlands Institute for Sound and Vision are the product of mergers. Practically, however, such an integration causes ontological challenges. This is exemplified by *Delta Finale*: four entries to it exist in Sound and Vision's catalogue, as remnants of constituent collections which are left invisible. This complicates access and obfuscates their provenance, limiting the manifestation of dialectics in collective cognition and hence the potential for environmental processes and politics. In order to maintain ontological robustness and to make audiovisual discourses manifest, attempts to integrate collections, including aggregate online platforms, must observe lower levels and specific environments and leave traces visible (Anger, 2024).

Besides this 'vertical' dimension across levels, there is also a 'horizontal' dimension of agents traversing different environments, which broadens the knowledge base, just as multiple coexisting archives can do. It opens up the possibility to think of an extended cultural ecology with multiple culture cores. While based on a different premise, I see this related to Caswell's (2014) decentralized network approach. But instead of conceptualizing it as memoryscapes, the coordination of the dimensions is better served by an ontologically grounded 'deltaic curating of dialectics'.

Memory in the individual interacts with other cognitive functions; archives may similarly serve 'collective cognition'. This implies an epistemological shift, which needs direction through 'deltaic curating of dialectics'. 'Deltaic' is characterized by the river meeting the sea, where saltwater and freshwater mix, affected by the tides and other cycles. But the delta is also characterized by multiple streams, interconnecting but still directional, which rephrases Hessler's 'tidalectic curating'. Above all, the delta is a living environment. Deltaic curating calls for an encounter between different forces and forms of life, which can be co-existing, complementary, symbiotic or opposed.

Deltaic curating aligns with the ecosystem of which it is part, and acknowledges its position in historical developments, while actualizing discourses. Deltaic curating of archival films goes beyond knowledge management and access. It provides an epistemological horizon that fosters multiplicity and dialectics. It enables evaluation, envisioning and intervention in a continuous cycle. These may serve as prompts for curators, researchers and planners to keep track of discursive positions and dialectics that invite building on ontological complexity.

Figure 1: Poster for the film *Plan Delta* (Visser, 1989). Collection Eye Filmmuseum.

Figure 2: Cover of the book *We stinken er in . . .!* (Van Hillo & Kielich, 1971a, 1971b). Van Holkema & Warendorf (Public domain).

Figure 3: Poster for the film *Armoede van de Toekomst* (Entrop, Hopmans, & Siegmund, 1981). Collection Eye Filmmuseum.

General References

Algemeen Dagblad (1989, 1 September). 'Zeeuws Filmfestival'. *Algemeen Dagblad*, 28.

Algemeen Handelsblad (1968, 18 December). 'Weg vrij voor tweede kernenergiecentrale: Péchiney vestigt zich in Zeeland.' *Algemeen Handelsblad,* 20. Available: www.delpher.nl (accessed 12 May 2025).

Anger, J. (2024). *Towards a Film Theory from Below: Archival Film and the Aesthetics of the Crack-Up* (New York: Bloomsbury).

Apaydin, V. (ed., 2020). *Critical Perspectives on Cultural Memory and Heritage: Construction, Transformation and Destruction* (London: University College London Press).

Appadurai, A. (2010). 'How Histories Make Geographies: Circulation and Context in a Global Perspective', *The Journal of Transcultural Studies*, 1(1), 4-13.

Assmann, J. (1995 [1988]). 'Collective Memory and Cultural Identity', *New German Critique, 65*, 125-133.

Bateson, G. (1987 [1972]). *Steps to an Ecology of Mind: Collected Essays in Anthropology, Psychiatry, Evolution, and Epistemology* (Northvale, NJ/London: Jason Aronson).

Beeld&Geluid (2025). *Collectie (Catalogus)*. Available: https://zoeken.beeldengeluid.nl (accessed 12 May 2025).

Beeld&Geluid (2025a). *Collectie (Catalogus).* 'Programma: Delta Fase 1'. [PID: URN:NBN:NL:IN:20-PGM3871968, ResourceID: 2101608130111029131]. Available: https://zoeken.beeldengeluid.nl (accessed 12 May 2025).

Beeld&Geluid (2025a). *Collectie (Catalogus).* 'Programma: Delta Finale 1988'. PID: URN:NBN:NL:IN:20-PGM78584. Available: https://zoeken.beeldengeluid.nl (accessed 12 May 2025).

Beeld&Geluid (2025b). *Collectie (Catalogus).* 'Programma: Delta Finale. Series: Bedrijfsfilms Baggerbedrijf van Oord: Delta Finale.' PID: URN:NBN:NL:IN:20-PGM3271452. Available: https://zoeken.beeldengeluid.nl (accessed 12 May 2025).

Beeld&Geluid (2025c). *Collectie (Catalogus).* 'Programma: Delta Finale. Series: Bedrijfsfilms ENCI B.V.: Delta Finale.' PID: URN:NBN:NL:IN:20-PGMGSV200618270LH. Available: https://zoeken.beeldengeluid.nl (accessed 12 May 2025).

Beeld&Geluid (2025d). *Collectie (Catalogus).* 'Programma: VNC Delta Finale. Series: Bedrijfsfilms ENCI B.V.' PID: URN:NBN:NL:IN:20-PGMGSV200618269ML. Available: https://zoeken.beeldengeluid.nl (accessed 12 May 2025).

Bhaduri, S. (2020). *Polycoloniality: European Transactions with Bengal from the 13th to the 19th Century* (New Delhi: Bloomsbury).

Bregman, R. (2023). *70 jaar na de Watersnoodramp is onze strijd tegen het water niet voorbij – integendeel.* De Correspondent. Available: https://decorrespondent.nl/14177/70-jaar-na-de-watersnoodramp-is-onze-strijd-tegen-het-water-niet-voorbij-integendeel/d48ee983-43ca-03f4-19bd-35622aa8e411 (accessed 1 December 2024).

Brunow, D. (2017). 'Curating Access to Audiovisual Heritage: Cultural Memory and Diversity in European Film Archives', *Image [&] Narrative*, 18(1), 97-110.

Caswell, M. (2014). 'Inventing New Archival Imaginaries: Theoretical Foundations for Identity-Based

Community Archives', in D. Daniel & A. Amalia & S. Levi (eds.), *Identity Palimpsets: Archiving Ethnicity in the U.S. and Canada*, pp. 35-58 (Sacramento, CA: Litwin Books).

Cook, T. (2013). 'Evidence, Memory, Identity, and Community: Four Shifting Archival Paradigms', *Archival Science, 13*, 95-120.

Croombs, M. (2019). 'In the Wake of Militant Cinema: Challenges for Film Studies', *Discourse, 41*(1), 68-89.

De Bevelander (1999, 1 September). 'Premières en sterren op filmfestival'. *De Bevelander*, , 1. Available: www.tijdschriftenbankzeeland.nl (accessed 12 May 2025).

De Jong, A. (1984). 'Tien jaar Gouden Delta in boekvorm vastgelegd', *Nieuw Zeeland*, 6-7, 16. Available: www.tijdschriftenbankzeeland.nl (accessed 12 May 2025).

De Ruiter, F. (1976, 20 November). 'Knokken tegen kernenergie, een document'. *NRC Handelsblad*, , 3. Available: www.delpher.nl (accessed 12 May 2025).

De Ruiter, F. (1985, 31 January). 'Waterstaat speurt naar buitenlandse orders'. *NRC Handelsblad*, , 2. Available: www.delpher.nl (accessed 12 May 2025).

De Stem (1970, 21 October). 'Plattelandsvrouwen: Studiedag over milieubehoud-zelfbehoud'. *De Stem*, , 3. Available: https://krantenbankzeeland.nl (accessed 12 May 2025).

De Stem (1971, 8 January). 'Vervolg op 'We Stinken erin': Van Hillo gebruikt Biesbosch als bewijs en illustratie'. *De Stem*,, 5. Available: https://krantenbankzeeland.nl (accessed 12 May 2025).

De Volkskrant (1986, 4 October). 'Plan Delta wijst op menselijke hoogmoed.' *De Volkskrant*, , 37.

De Waarheid (1981, 2 September). 'Dodewaard gaat dicht!', *De Waarheid*, 1981, 4. Available: www.delpher.nl (accessed 12 May 2025).

Delpher (2025). *KB Nationale Bibliotheek*. Available: https://www.delpher.nl (accessed 12 May 2025).

Erll, A. (2010). 'Cultural Memory Studies: An Introduction' in: A. Errl & A. Nünning (eds.), *Cultural Memory Studies: An International and Interdisciplinary Handbook*, 1-15 (Berlin/New York: De Gruyter).

EUscreen (2025). *Stichting EUscreen* (Utrecht). Available: https://euscreen.eu (accessed 12 May 2025).

Eye (2025), *CE: Colletion Eye (Catalogus).* https://www.eyefilm.nl/en/collection/eye-catalogue (accessed 12 May 2025).

Eye (2025a), *CE: Colletion Eye (Catalogus).* 'Delta Fase 1'. Record ID: FLM15036. Available: https://www.eyefilm.nl/en/collection/eye-catalogue (accessed 12 May 2025).

Foucault, M. (2002 [1969]). 'The Historical *A Priori* and the Archive', in M. Foucault, *Archaeology of Knowledge*, 142-148 (London: Routledge).

Gowricharn, R. (2024). *The Girmitiya Peasants in Suriname: Agrarian and Economic Transformations in a Plantation Society* (Cham: Springer).

Halbwachs, M. (1925). *Les cadres sociaux de la mémoire* (Paris: Librairie Félix Alcan).

Halbwachs, M. (1980 [1950]). 'Space and the Collective Memory', in M. Halbwachs, *The Collective Memory*, 128-156. (New York: Harper Colophon Books).

Harris, V. (2012). 'Genres of the Trace: Memory, Archives and Trouble', *Archives and Manuscripts*, 40(3), 147-157.

Hedstrom, M. (2010). 'Archives and Collective Memory: More than a Metaphor, Less than an Analogy', in T. Eastwood & H. MacNeil (eds.), *Currents of Archival Thinking*, 163-179 (Santa Barbara: Libraries Unlimited).

Hessler, S. (2020). 'Tidalectic Curating', *Journal of Curatorial Studies*, 9(2), 249-270.

Iacob, B. (2023). 'Curating as Technique versus Curating as Mediation: A Proposal for a More Lucid Understanding of Curatorial Practices', in B. Iacob & T. Pop (eds.), *Cercetări actuale în formele și mediile expresive: Teorii și practici artistice*, 21-44 (Cluj: Presa Universitară Clujeană).

IDFA. (2024). *Kamar Ahmad Simon*. Available: www.idfa.nl/person/2453525f-d7b3-4770-a476-5abc0a01688f/kamar-ahmad-simon (accessed 25 November 2024).

IFFR. (2024). *Stichting International Film Festival Rotterdam*. Available: https://iffr.com/en (accessed 20 December 2024).

IISG (2024). *Greenpeace film clips.* Available: https://iisg.nl/collections/greenpeace/film-nl.php (accessed 9 December 2024).

IVN (2024b). *Om het laatste groen* (film). Available: https://ivngeschiedenis.nl/IVN_6069/Achtergrond.html (accessed 25 November 2024).

Jager, I. (1989, 30 January). 'Lage landen gefilmd na nieuwe watersnood'. *De Volkskrant*, 7.

Jimerson, R. (2009). *Archives Power: Memory, Accountability, and Social Justice*. Chicago: Society of American Archivists, 2009.

Jongejan, M. (1991, 1 March). 'Het Zeeuws Filmarchief', *Nehalennia*, 33-36. Available: https://tijdschriftenbankzeeland.nl/issue/neh/1991-03-01/edition/null/page/35?query= (accessed 1 December 2024).

Kerker, R. (2015). *Delta Finale, YouTube.* Available: www.youtube.com/watch?v=sAWQ8TRYfao (accessed 12 December 2024).

Kok, A. (1989, 3 February). 'Vaderlandse kopstukken mijden festival Rotterdam'. *Algemeen Dagblad*, 21.

Kotterer, F. (1989, 1 February). 'Pim de la Parra's eigen hommage aan Huub Bals'. *Het Parool*, 23.

Krantenbank Zeeland (2025). *ZB Bibliotheek van Zeeland*. Available: https://krantenbankzeeland.nl (accessed 12 May 2025).

Laka (2024). *Eerste Dodewaard-gaat-dicht! blokkade-actie.* Available: https://www.kernenergieinnederland.nl/node/204 (accessed 1 December 2024).

Löw, M. (2013). 'The City as Experiential Space: The Production of Shared Meaning', *International Journal of Urban and Regional Research*, 37(3), 894-908.

Luhmann, N. (1997). 'Globalization or World Society: How to Conceive of Modern Society?' *International Review of Sociology,* 7(1), 67-79.

Marks, L. (2004). 'The Ethical Presenter: Or How to Have Good Arguments over Dinner', *The Moving Image,* 4(1), 35-47.

McDowell, S. (2008). 'Heritage, Memory, and Identity', in B. Graham & P. Howard (eds.), *The Ashgate Research Companion to Heritage and Identity*, 37-53 (London: Routledge).

Mdftrasher (2023). *Delta Fase 1 1996, YouTube.* [Video and comments section]. Available: www.youtube.com/watch?v=Tr46oVi-74c (accessed 20 December 2024).

Mediapolis. (2024). *Mediapolis: A Journal of Cities and Culture*. Available: www.mediapolisjournal.com (accessed 12 May 2025).

Media Suite. (2025). *Clariah Media Suite*. Available: https://mediasuite.clariah.nl (accessed 12 May 2025).

Molendatabase. (2023). *Middelburg, Zeeland: Getijdenmolen bij de Nieuwe Haven*. Available: www.molendatabase.nl/molens/ten-bruggencate-nr-11215 (accessed 20 December 2024).

Monticelli, R. (2019). '*Under the Skin*: Science Fiction as Posthuman Cultural Ecology', *Between*, 9(17), 2-24.

Newell, J., Robin, L., & Wehner, K. (eds., 2017). *Curating the Future: Museums, Communities and Climate Change* (London: Routledge).

NRC Handelsblad (1992). 'Filmfestival Zeeland.' *NRC Handelsblad*, 3 September 1992, 6. Available: www.delpher.nl (accessed 12 May 2025).

Paalman, F. (2011). *Cinematic Rotterdam: The Times and Tides of a Modern City* (Rotterdam: 010 Publishers).

Paumen, M., & De Ruiter, F. (1984). *'De gouden pijlers van de Oosterschelde.' NRC Handelsblad*, 23 June 1984, 3. Available: www.delpher.nl (accessed 12 May 2025).

Prakash, O. (1985). *The Dutch East India Company and the Economy of Bengal, 1630-1720* (Princeton, NJ: Princeton University Press).

Provinciale Zeeuwsche Courant (1981, 29 January). 'Jongerenjournaal over kernenergie: Met filmbeelden demonstratie Borssele'. *Provinciale Zeeuwsche Courant*, 9. Available: https://krantenbankzeeland.nl (accessed 12 May 2025).

Provinciale Zeeuwse Courant (1990, 7 September). 'Filmfestival Zeeland.' *Provinciale Zeeuwse Courant*, 32. Available: https://krantenbankzeeland.nl (accessed 12 May 2025).

Slavenburg, P. (1989, 26 January). 'Plan Delta, een film met een decor van tien miljard gulden.' *Het Vrije Volk*, 7. Available: www.delpher.nl (accessed 12 May 2025).

Steward, J. (1976 [1955]). *Theory of Culture Change: The Methodology of Multilinear Evolution* (Urbana, IL: University of Illinois Press).

Thomassen, W. (1975, 13 December). 'Baggeraars pionieren wat af in Bangladesj.' *Algemeen Dagblad*, 15. Available: www.delpher.nl (accessed 12 May 2025).

Tijdschriftenbank Zeeland (2025). *ZB Bibliotheek van Zeeland*. Available: www.tijdschriftenbankzeeland.nl (accessed 12 May 2025).

Touran, N. (2007-2025). 'Historical Nuclear Energy Films.' Available: https://whatisnuclear.com/old-videos.html (accessed 20 November 2024).

Van Bueren, P. (1989, 1 February). 'Oosterscheldedam als inspiratie.' *De Volkskrant*, 13. Available: www.delpher.nl (accessed 12 May 2025).

Van Damme, J. (1989, 8 September). 'Alsof je niet meer bij de wereld hoorde.' *Provinciale Zeeuwse Courant*, 13. Available: www.delpher.nl (accessed 12 May 2025).

Van der Peijl, D. (1985, 20 March). 'De Deltawerken van Bangladesh'. *NRC Handelsblad*, 5. Available: www.delpher.nl (accessed 12 May 2025).

Van Hillo, J., & Kielich, W. (1971a). *We stinken er in . . .! Een momentopname van ons leven op een vulkaan* (Bussum: Van Holkema & Warendorf).

Van Hillo, J., & Kielich, W. (1971b). *We stinken er in . . .! Deel 2: Het uur van de waarheid. Een televisiedocumentaire* (Bussum: Van Holkema & Warendorf).

Veld, A. (1989). 'Plan Delta: 'Mijn stijl is dat er geen stijlvastheid is", *Film- en tv-maker*, 291, 18-21.

Veltman, F., & Dogger, H. (2015, 2 September). *Groeten uit Renesse.* Andere Tijden. VPRO/NTR. Available: https://anderetijden.nl/aflevering/1/Groeten-uit-Renesse (accessed 7 July 2024).

Unie van Waterschappen (2024). *Waterschapskaart.* Available: https://unievanwaterschappen.nl/publicaties/waterschapskaart/ (accessed 25 November 2024).

Wegener Sleeswijk, A. (2024). Auctions and the Distribution of Silks in the Eighteenth-Century United Provinces', in P. Vernus, M. Martini, M. & T. Hashino (eds.). *A Global History of Silk: Trade and Production from the 16th to the Mid-20th Century*, 27-55 (Cham: Springer).

Wij zijn de stad (2018). *Wij zijn De Stad, Middelburg.* Available: https://wijzijndestad.com/verhalen/middelburgers-jan-willem-antheunisse (accessed 25 November 2024).

Wikipedia (2023). *Borssele Nuclear Power Station.* Available: https://en.wikipedia.org/wiki/Borssele_Nuclear_Power_Station (accessed 12 May 2025).

WWF (2024). *Zoet en Zout komen samen.* Available: https://www.wwf.nl/wat-we-doen/waar-zijn-we-actief/nederland/delta (accessed 7 July 2024).

Zapf, H. (2021). 'Cultural Ecology, the Environmental Humanities, and the Ecological Archives of Literature', *Colloquia Germanica – Themenheft: Ecological Archives*, 53(2-3), 121-140.

Zeeuws Archief (2024). *Verenigde Oostindische Compagnie kamer Zeeland*. Available: www.zeeuwsarchief.nl/zeeuwse-verhalen/verenigde-oostindische-compagnie-kamer-zeeland (accessed 20 December 2024).

Zeeuwse Ankers (2024). *Protestbewegingen tegen afsluiting van de Oosterschelde.* Available: www.zeeuwseankers.nl/verhaal/protestbewegingen-tegen-afsluiting-van-de-oosterschelde (accessed 29 November 2024).

Zeeuws Archief (2002). 676 Zeeuwse Milieufederatie (ZMF) en voorgangers, 1966-2002; 5.1.1. Periodiek Bulletin VMZ/De Gouden Delta. Zeeuws Archief, Middelburg.

Zeeuws Filmarchief. (1992). *Catalogus: Zeeland in Film en Video*. Middelburg: Zeeuws Filmarchief/Zeeuws Documentatiecentrum.

Films

Mullens, A., & Mullens, W. (Directors, 1905). *De mésaventure van een Fransch heertje zonder pantalon aan het strand te Zandvoort* [Film]. Albert Frères [Collection: Eye]. Available: https://filmdatabase.eyefilm.nl/collectie/filmgeschiedenis/film/de-m%C3%A9saventure-van-een-fransch-heertje-zonder-pantalon-aan-het (accessed 12 May 2025).

Atomic Energy Commission (Commissioner, 1969). *Project Salt Vault* [Film]. Oak Ridge National Laboratory. [Collection: Eye, ID: FLM54990].

Batchelor, J. (Director, 1961). *Dam the Delta / Nederland – Deltaland* [Film]. J. Fernhout (Producer). [Collection: Eye, ID: FLM45637 and NISV, ResourceID: 2101608130111036931].

Breijer, Ch. (Director, 1969). *Delta Ypsilon / Sluiting van het Volkerak* [Film). Multifilm [Collection: NISV, digital file code: DELTA_YPSILON-DPX5672306R.tar].

Breijer, Ch. (Director, 1974). *Spuisluizen in het haringvliet* / Hoofdkraan van Nederland [Film]. Multifilm [Collection: NISV, ResourceID: 2101608050034391431].

Burcksen, J., & Herblot, R. (Directors, 1983). *Dutch Delta* [Film]. Mundofilm. [Collection: Eye, ID: FLM140393; NISV, ResourceID: 2101608130110589931. Available: https://mediasuite.clariah.nl (accessed 12 May 2025).

Burcksen, J., & Herblot, R. (Directors, 1985). *Delta Finale* [Film]. Mundofilm.

Burcksen, J., & Herblot, R. (Directors, 1986). *Delta Finale* [Film]. Mundofilm.

Burcksen, J., & Herblot, R. (Directors, 1988). *Delta Finale* [Film]. Mundofilm. [Collection: NISV, ResourceID: 2101608050034372131].

Entrop, B., Hopmans, A., & Siegmund, M. (Directors, 1981). *Armoede van de toekomst* (jongerenjournaal nr. 2) [Film]. Solidariteitsfilms. [Collection: Eye, ID: FLM4015].

Entrop, B., & Hopmans, A., & Siegmund, M. (Directors, 1981). *Speelgoed voor later* (jongerenjournaal nr. 3) [Film]. Solidariteitsfilms. [Collection: Eye, ID: FLM62640].

Erends, R. (Director, 1960). *Atomos*. [Collection: Eye, ID: FLM4438].

Fernhout, J. (Director, 1968). *Delta Data* [Film]. J. Fernhout (Producer). [Collection: NISV, ResourceID: 2101608130112015331]. Available: https://mediasuite.clariah.nl (accessed 12 May 2025).

Gladitz, N. (Director, 1976). *Lieber heute aktiv als morgen radioaktiv* [Film]. Gladitz-Teldok Produktion [Collection: IISG, Cineclub Vrijheidsfilms, BG F2/211-213].

Haanstra, B. (Director, 1961). *Delta Fase 1* [Film]. Bert Haanstra Films B.V. [Collection: NIBG, Eye].

Haanstra, B. (Director, 1966). *The Voice of the Water* [Film]. Bert Haanstra Films B.V.

Heerkens, N. (Director, 1977). *Demonstratie Kalkar, 1977* [Film]. Noud Heerkens. [Collection: Eye].

Holthuis, G. (Director, 1981). *Fiction* [Film]. Filmstad [Collection: Eye, ID: FLM20785].

Hopmans, A. (Director, 1980). *Erbij horen* (jongerenjournaal nr. 1) [Film]. Solidariteitsfilms. [Collection: Eye, ID: FLM18994].

Hoving, H. (Director, 1958-1961). *De sluiting van het veerse gat* [Film]. Multifilm. [Collection: NISV, Acte 1 (1958-07-01): ResourceID: 2101608050034348131; Acte 2 (1961-08-01): 2101608050034347731; Acte 3 (1961-04-24): 2101608050034347131]. Available: https://mediasuite.clariah.nl (accessed 12 May 2025).

Islam, M. (Director, 1997). *Dukhai* [Film]. Monon Chalachitra. Digital Media Lab.

IVN. (Producer, 1965). *Om het laatste groen* [Film]. IVN.

Jansen, J., & Elzinga, R. (Directors, 1981). *Dodewaard gaat dicht: blokkade van de kerncentrale Dodewaard* [Film]. Dodewaard Gaat Dicht! [Collection: IISG, Cineclub Vrijheidsfilms, BG F2/125-126].

Keller, D. (Director, 1975). *Lovejoy's Nuclear War* [Film]. Green Mountain Post Films. [Collection: IISG, Cineclub Vrijheidsfilms: BG F2/71].

Rerelease, ZinnEdProject (2016). Available: www.youtube.com/watch?v=0nRnDfVV72Q (accessed 12 May 2025).
Polygoon (Producer, 1969-32). *Polygoon Journaal: Opening Heinenoordtunnel*. [Newsreel] [Collection: NISV, ResourceID: 2101608140120232131]. Available: www.youtube.com/watch?v=xH-wsS8tpZAg&t=0s (accessed 12 May 2025).
Polygoon (Producer, 1979-16). *Polygoon Journaal: Kerncentrale in Borssele*. [Newsreel] [Collection: NISV, ResourceID: 2101608140120282631. Available: www.youtube.com/watch?v=ux-ySoJtLwc, (Accessed 3 December 2024).
Reiné, R. (Director, 2015). *Admiral* [Film]. Farmhous Film & TV.
Rutten, G. (Director, 1934). *Dead Water* [Film]. Nederlandsche Filmgemeenschap.
Simon, K. (Director, 2012). *Are You Listening!* [Film]. Studio BEGINNING.
Simon, K. (Director, 2021). *Day after...* [Film]. Studio BEGINNING.
Socialistiese Partij. (1976) [Film]. Socialistiese Partij [Coll. Rotterdam City Archive: BB-1807].
Visser, B. (Director, 1989). *Plan Delta* [Film]. Neon Film. [Collection: Eye, ID: FLM53154].

Television and webvideos

AVRO (Broadcaster, 1970, 14 July). *Televizier Magazine* [Television programme]. AVRO Television. [Collection: NISV, ResourceID: 2101608060046921831]. Available: https://mediasuite.clariah.nl (accessed 12 May 2025).
AVRO (Broadcaster, 1979, 5 April). *Televizier Magazine* [Television programme]. AVRO Television. [Collection: NISV, ResourcedID: 2101608060047021431]. Available: https://mediasuite.clariah.nl (accessed 12 May 2025).
EBU (Broadcasting service, 1965). *Internationale nieuwsuitwisseling: Item 8. Overstroming in Bangladesh* [Television news] European Broadcasting Union [Collection: NISV, ResourceID: 2101608070056968831]. Available: https://mediasuite.clariah.nl (accessed 12 May 2025).
EBU (Broadcasting service, 1984, 25 August). *Internationale nieuwsuitwisseling: Rode Kruis biedt hulp aan slachtoffers van overstroming Bangladesh* [Television news]. European Broadcasting Union [Collection: NISV, ResourceID: 2101608070057410731]. Available: https://mediasuite.clariah.nl (accessed 12 May 2025).
Fijnvandraat, J. (Director), & Le Poole, H. (Editor-in-chief, 2003, 1 February). *Holland Sings – Commemoration of the flood-disaster / Nederland herdenkt en zingt: Herdenking Watersnoodramp* [TV programme]. J. de Haan (Producer). EO Television. [Collection: NISV, PID: URN:NBN:NL:IN:20-PGM158321], Available: https://euscreen.eu (accessed 12 May 2025).
Haanstra, B. (Director), & Van Fucht, M. (Series director), & Bijlholt, P. (Editor-in-chief KRO, 1996, 8 May). Nederland (1982) en Delta Phase 1 (1962) (4th episode) [TV series episode]. In Y. Vluggen (Producer), *De films van Bert Haanstra*. Bert Haanstra Films B.V., KRO Televisie. [Collection: NISV, ResourceID: 2101608130115450731]. Available: https://mediasuite.clariah.nl (accessed 12 May 2025).

Hulscher, H. (Director), & Cuypers, L. (Composer, 1978, 10 January). *Zeeland Suite* [Television programme]. NOS Television. [Collection: NISV, ResourceID: 2101608050036985531]. Available: https://mediasuite.clariah.nl (accessed 12 May 2025).

Kerker, R. (2015, March 18). *Delta Finale* [Video]. YouTube. Available: www.youtube.com/watch?v=sAWQ8TRYfao (accessed 12 May 2025).

KRO (Broadcaster, 1973, 12 May). *Brandpunt* [Television programme]. KRO Television [Collection: NISV, ResourceID: 2101608060046816431].

KRO (Broadcaster, 1974, 13 January). *Brandpunt* [Television programme]. KRO Television [Collection: NISV, ResourceID: 2101608060046818331]. Available: https://mediasuite.clariah.nl (accessed 12 May 2025).

Mdftrasher. (2023, 6 February). *Delta Fase 1 1996* [Video]. YouTube. Available: https://www.youtube.com/watch?v=Tr46oVi-74c, (accessed 12 May 2025).

NCRV (Broadcaster, 1974, 26 January). *Farce Majeure*. NCRV Television [Collection: NISV, ResourceID: 2101608120099363131]. Available: https://mediasuite.clariah.nl (accessed 12 May 2025).

NCRV (Broadcaster, 1973, 15 May). *Hier en Nu* [Television programme]. NCRV Television. [Collection: NISV, ResourceID: 2101608060047018331].

NCRV (Broadcaster, 1984, 16 November). *Hier en Nu* [Television programme]. NCRV Television. [Collection: NISV, ResourceID: 2101608060047100731]. Available: https://mediasuite.clariah.nl (accessed 12 May 2025).

NCRV (Broadcaster, 1984, 21 December). *Hier en Nu* [Television programme]. NCRV Television. [Collection: NISV, ResourceID: 2101608050038386531]. Available: https://mediasuite.clariah.nl (accessed 12 May 2025).

NCRV (Broadcaster, 1989, 1 September). *Vijfenvijftig jaar NCRV: Hoogtepunten uit 55 jaar NCRV-uitzendingen* [Television programme]. NCRV Television. [Collection: NISV, ResourcedID: 2101608050034524831]. Available: https://mediasuite.clariah.nl (accessed 12 May 2025).

NCRV (Broadcaster, 1994, 2 December). *O, ja! De NCRV in beeld: Jan van Hillo* [Television programme]. NCRV Television. [Collection: NISV, ResourceID: 2101608120102876431]. Available: https://mediasuite.clariah.nl (accessed 12 May 2025).

NOS (Broadcaster, 1979, 1 April). *Journaal* [TV news]. NOS Television. [Collection: NISV, ResourceID: 2101608080063522331]. Available: https://mediasuite.clariah.nl (accessed 12 May 2025).

NOS (Broadcaster, 1979, 2 April). *Journaal* [TV news]. NOS Television. [Collection: NISV, ResourceID: 2101608080063522031]. Available: https://mediasuite.clariah.nl (accessed 12 May 2025).

NOS (Broadcaster, 1979, 7 April). *Journaal* [Television news]. NOS Television. [Collection: NISV, ResourceID: 2101608080063522431]. Available: https://mediasuite.clariah.nl (accessed 12 May 2025).

NOS (Broadcaster, 1983, 9 November). *Van Gewest tot Gewest (Gewest Kort): Vereniging Milieuhygiëne Zeeland: de Gouden Delta.* [Television programme]. NOS Television. [Collection: NISV, ResourceID: 2101608060047391931]. Available: https://mediasuite.clariah.nl (accessed 12 May 2025).

NOS (Broadcaster, 1988, 11 September). *Journaal* [TV news]. NOS Television [Collection: NISV, ResourceID: 2101608080063933031]. Available: https://mediasuite.clariah.nl (accessed 12 May 2025).

NOS (Broadcaster, 1988, 13 September). *Journaal* [TV news]. NOS Television. [Collection: NISV, ResourceID: 2101608040033449631]. Available: https://mediasuite.clariah.nl (accessed 12 May 2025).

Omroep Zeeland. [2001]. (2020, 10 April). *Trugkieke – Kerncentrale Borssele, 1973 / 74* [Video]. Omroep Zeeland. Available: www.youtube.com/watch?v=RBuxNA3fKa4&list=PLE7554272918B150A&index=174 (accessed 12 May 2025).

Omroep Zeeland. [2001]. (2020, 25 May). *Trugkieke – Zeeland in de Hilversumse kijker VI* [Video]. Omroep Zeeland. Available: www.youtube.com/watch?v=oI1q0Fc3pAE (accessed 12 May 2025).

Omroep Zeeland. [2003]. (2020, 28 October). *Trugkieke – Deltaplan: Bouw Stormvloedkering* [Video]. Omroep Zeeland. Available: www.youtube.com/watch?v=o7TOH661mHQ (accessed 12 May 2025).

Omroep Zeeland. (2022, 6 October). *25 jaar Trugkieke* [Video]. Omroep Zeeland. Available: www.omroepzeeland.nl/video/25-jaar-Trugkieke-ZNO220927KU/25-jaar-trugkieke (accessed 12 May 2025).

Schaper, J. (Director, 1966, 26 May). *Het proces Renesse* [Film]. J. van Hillo (Producer), NCRV Television. [Collection: Stadsarchief Rotterdam]. Available: https://anderetijden.nl/aflevering/1/Groeten-uit-Renesse (accessed 12 May 2025).

Spoor, R. (Director, 1986, 4 October). *Storm Surge Barrier Eastern Schelde / Stormvloedkering Oosterschelde* [TV programme]. Ch. Te Nuyl (Producer). NOS Television. [Collection: NISV, ResourceID: 2101608050035350331]. Available: https://euscreen.eu/collection-search-results/?search=delta%20netherlands (accessed 12 May 2025).

TELEAC (Broadcaster, 1976, 22 February). De delta (Episode 3) [TV series episode]. *Bedreigde landschappen*. TELEAC. [Collection: NISV, ResourceID: 2101608050039105331]. Available: https://mediasuite.clariah.nl (accessed 12 May 2025).

TROS (Broadcaster, 1974, 23 January). Kernenergie en stralingsgevaar [TV series episode]. *Op uw gezondheid* [TV series]. [Collection: ResourceID, 2101608060043559931]. Available: https://mediasuite.clariah.nl (accessed 12 May 2025).

TROS (Broadcaster, 1979, 7 April). *Aktua* [TV programme]. TROS Television. [Collection: NISV, ResourceID: 2101608060047631431]. Available: https://mediasuite.clariah.nl (accessed 12 May 2025).

TROS (Broadcaster, 1982, 5 October). *Aktua – Item 4. De Oosterscheldedam* [TV programme]. TROS Television. [Collection: NISV, ResourceID: 2101608060047601931]. Available: https://mediasuite.clariah.nl (accessed 12 May 2025).

TROS (Broadcaster, 1988, 6 September). *Aktua – Item 3. Hulpverlening tbv Bangladesh* [TV programme]. TROS Television. [Collection: NISV, ResourceID: 2101608060047609031]. Available: https://mediasuite.clariah.nl (accessed 12 May 2025).

Truyman, A. (Director, 1971, 19 January). *Het land van de dwazen* (Episode 2). [TV series]. [Collection: NISV, ResourceID: 2101608050038288431]. Available: https://mediasuite.clariah.nl (accessed 12 May 2025).

Van der Velde, W., & Smits, W. (Directors, 1961, 8 November). *Polders voor industrie* [TV programme]. VPRO Television. [Collection Rotterdam City Archive: BB-0730].

Van Hillo, J. (Director, 1970, 20 September). *Mijlpaal in het Haringvliet* Television programme]. NCRV Television. [Collection: NISV]. Available: https://mediasuite.clariah.nl (accessed 12 May 2025).

Van Hillo, J. (Director, 1970, 26 October and 11 November). Een momentopname van ons leven op een vulkaan (Episode 1) [TV series episode]. *We stinken erin...!* NCRV Television [Collection: NISV, ResourceID: 2101608060043719131]. Available: https://mediasuite.clariah.nl (accessed 12 May 2025).

Van Hillo, J. (Director, 1971, 29 January). Het uur van de waarheid (Episode 2) [TV series episode]. *We stinken erin...!* NCRV Television [Collection: NISV, ResourceID: 2101608170159263631]. Available: https://mediasuite.clariah.nl (accessed 12 May 2025).

Van Hillo, J. (Director, 1971, 14 May). 'n Goede raad is duur... (Episode 3) [TV series episode]. *We stinken erin...!* NCRV Television [Collection: NISV, ResourceID: 2101608060043755431]. Available: https://mediasuite.clariah.nl (accessed 12 May 2025).

Van Hillo, J. (Director, 1971, 1 October). Vermenigvuldigen (Episode 4) [TV series episode]. *We stinken erin...!* NCRV Television [Collection: NISV, ResourceID: 2101608060044142031], Available: https://mediasuite.clariah.nl (accessed 12 May 2025).

Van Hillo, J. (Director, 1971, 22 December). De Toekomstschok (Episode 5) [TV series episode]. *We stinken erin...!* NCRV Television [Collection: NISV, ResourceID: 2101608060043791031], Available: https://mediasuite.clariah.nl (accessed 12 May 2025).

VARA (Broadcaster, 1979, 3 April). *VARA-Visie* [Television programme]. VARA Television [Collection: NISV, ResourceID: 2101608060047755631]. Available: https://mediasuite.clariah.nl (accessed 12 May 2025).

Veronica (Broadcaster, 1988, 21 September). *Nieuwslijn* [TV programme]. VOO Television [Collection: NISV, ResourceID: 2101608060047836331]. Available: https://mediasuite.clariah.nl (accessed 12 May 2025).

VPRO (Broadcaster, 1980, 20 February), *Villa VPRO* (Television programme). VPRO Television. [Collection: NISV, ResourceID: 2101608100079583231]. Available: https://mediasuite.clariah.nl (accessed 12 May 2025).

VPRO (Broadcaster, 1984, 1 February), *Tape-TV: Aflevering 2* (Television programme). VPRO Television. [Collection: NISV, ResourceID: 2102311160358644631]. Available: https://mediasuite.clariah.nl (accessed 12 May 2025).

Wij zijn de Stad. (2017, 12 November). *Filmtheaters Middelburg in de jaren 80* [Video]. YouTube. Available: www.youtube.com/watch?v=UAzB-v4wpSyQ (accessed 18 December 2024).

Dialogue with Dirk Sijmons and Jord den Hollander

Dirk Sijmons, Jord den Hollander, Luca Iuorio and Sophia Arbara

***The Rhine from Lobith to the Sea*: 17 Years Later**

This dialogue was conducted as part of the closing screening of the Reporting the Delta itinerant film club. The event, which took place in June 2024 at the faculty of Architecture and the Built Environment at TU Delft, included a screening of two movies by Jord den Hollander, *Stroom: Collage van het collectieve geheugen (Stream: Collage of the collective memory*; 2005) and *De Rijn van Lobith tot aan zee 2007* (*The Rhine from Lobith to the Sea 2007*; 2008), followed by a discussion with Jord van Hollander and Dirk Sijmons. *Stroom* is a film about the changing river landscape at Nijmegen based on hundreds of archive films. *De Rijn van Lobith tot aan zee 2007* is a remake of the homonymous film by A. Ochse and A. M. van der Wel, produced in 1922. In the film, architect and filmmaker Jord den Hollander and national landscape advisor and landscape architect Dirk Sijmons retrace the journey by boat over the Rhine, reflecting on the changing landscape, 85 years later. The Reporting the Delta project discussed with Jord den Hollander and Dirk Sijmons – seventeen years after the film's remake and 102 years after the original film – the past, present and future delta transformations and the role of films and archival research in urbanism.

Deltaic landscape transformations

Sophia Arbara: Reflecting on the century since the original film, *De Rijn van Lobith tot aan zee* (*The Rhine from Lobith to the Sea*; 1922), and the seventeen years since its re-release, you mention that there have been unimaginable changes but also elements that have remained consistent. Could you elaborate on these changes, particularly since 2007? You discuss a shift from a production-driven to a consumption-driven landscape in the 2008 film. Do you feel we are still in an era of consumption, or have we entered a new phase?

Dirk Sijmons: The question of whether we remain in a consumption-driven era has both a 'yes' and 'no' answer. Yes, because in a way the Rhine continues to serve as the artery for the country's hedonistic tendencies. The leisure industry, especially aquatic leisure, has grown exponentially since the film was made. But there's also a contrasting trend – a second wave of privatization in the Dutch countryside driven by intensified agriculture. The first privatization wave was marked by the disappearance of the public character of the landscape and a decrease of public rights of way and dirt roads. These were the entry points to the landscape. It was a period where the small scale of the landscape was lost; the relics from earlier periods of land reclamations were lost in the effort to modernize and upscale agriculture after World War II.

In the recent past, after Jord's film was made, the agricultural sector scaled up even more and intensified, excluding any people from the farmyards that don't belong there. It marked a gradual new privatization of the Dutch countryside. So yes, things have changed, and I think I was very accurate in my prognosis about the leisure sector in introducing the 'horsification' of the landscape: we have more horses now than before the era of the tractor – but it has not been such a dominant force of the Dutch countryside as I anticipated seventeen years ago.

Sophia: The detachment of the agricultural sector brings up a related question about the different activities present in the deltaic landscape. In the 1922 film by Ochse and Van der Wel, we see a more harmonious relationship between elements like cattle and waterways, but in the 2007 version, these activities appear more segmented. The port, for instance, now seems isolated from the city, which has almost lost its port city identity. You also just mentioned an increasing detachment of the agricultural sector. Are we witnessing a growing polarization between the city and the areas beyond? If so, how do these transformations impact the cohesion of the landscape?

Dirk: The Dutch landscape was never designed for aesthetic appeal, yet in a way that is also part of its beauty. From the farmers' perspective, this detachment originates from the fact that the landscape is being seen as their property and perceived by them as an agricultural industrial estate. From an urban perspective, the Dutch landscape is perceived more and more as what I call the 'Unmoving Arcadian backdrop': a rural escape from our dynamic city life. This Arcadian landscape has been evolving rapidly over the last twenty years. Back in the 1980s, when I worked for the state forestry service, we helped design the new production landscapes across the Netherlands. That line of work ended in the mid-80s and since then, the landscape has been shaped largely by autonomous forces in the liberalized market, coming from within the farming sector itself. There has been a spectacular scaling up of the size of farms, while the number of farms has decreased rapidly compared to twenty or thirty years ago. This tendency is still accelerating, sadly. In my view, agricultural innovation is headed in the wrong direction with farm consolidation and intensification. My office was in Utrecht, near the countryside, and when we started there were still a lot of farmers. A decade later, when we moved to Amersfoort, there were only a few left, covering the same surface. Capital intensity and knowledge intensity have gone up too. It is a spectacular change in just two decades, and one that I believe is problematic; we need more, not fewer, farmers.

Jord den Hollander: As a filmmaker, I was deeply interested in the psychological and mental aspects of how we perceive landscapes. Not just as production landscapes or leisure landscapes, but rather how the river and the whole delta are integral to nation building and national identity. In Ochse and Van der Wel's film in 1922, the government was aware of the importance of building a national consciousness and the rivers, the IJssel, the Waal and the Rijn, played a very important role in this process. There are writings that speak about the taming of the river. At the same time, in the nineteenth century, our awareness grew. The Rhine was seen as a 'mother' that nurtured those who came to live in the Netherlands. This connection between the river and the power of the nation is very strong, and reflected in many films which convey a sense of national pride of the Netherlands. This is reflected in how we can produce things that are ours; typical Dutch products, from Dutch cheese to cows, and the bricks we made, sourced from the clay and transported by the river. This theme of the river being almost the DNA of our country is still very much alive. You observe it in the new right-wing parties: the river is always used as a symbol for the bloodline of our production. It's a very important feature and it has political meaning as well.

Luca Iuorio: The word 'production' comes up often, both in the film and in our discussion. There seems to be a tension between this idea of production and the landscape's leisure aspects. You suggest that, in a way, the landscape is still being 'produced' and that this production brings the river and water even closer. There's a kind of beauty in that. In Ochse and Van der Wel's film, traditional natural beauty is contrasted later with a 'beauty' that emerges from production. Where do we find beauty in today's landscape, especially in this kind of new production? If it exists, what does that beauty look like?

Jord: People relate to the landscape in completely different ways. I think the beauty lies in going to the landscape, observing the river and becoming aware of its deep, unseen history, hidden due to its constant flow. The endless water cycle, vaporizing, moving from the sea to coast and land, transforming into clouds and rain; it's a reminder of continuity. The attachment that someone feels to such a landscape is part of the beauty. This kind of water landscape connects us to nature on a deeper level; we are aware that there have been thousands of years of continuity. At the same time, we start to see what the culture, our culture added to it: the industry, the culture of agriculture, the recreation; and how it's just a small part of a larger natural system. I remember that after Ochse and Van der Wel's film, there was a period of fifty years in which the river was used as a sewage system. Recreation became possible only when the river was cleaned. It did not just happen. First, the water was cleaned and then suddenly people realized that they could swim. At that time, the filth of Europe arrived in the Netherlands and we accepted that because it brought us profit. Today, we have a completely different view of the whole river system.

Luca: In the movie you discuss the constructed landscape of the delta and comment on how fortunate we are to live in this delta. Do you still think we are lucky to live in this constructed delta?

Dirk: Indeed, the delta is constructed all the way up to its capillaries. You could even say that the landscape of the west of the Netherlands is a prosthesis that allows us to live and work below sea level. When the film was made, I think we were in the early stages of the Room for the River program. By 2016, the last part of the Room for River project was executed. Since then, we have experienced a couple of instances where the water level in the river was very high. I was very happy to see a lot of people coming to see these new floodscapes. It was to some extent a resurrection, and beautiful to see: a chance to witness something that once would have incited fear, such as high water levels, now controlled and managed. I saw it in Nijmegen, near where the film begins. All of

Nijmegen came out to see it and that is when I felt very fortunate to have made a contribution to a project like this. While there is a feeling of control, we have to stay very cautious, because we are still basically vulnerable; and in terms of climate change I'm afraid we have seen nothing yet.

This was an interesting part of the Room for the River project. Culturally, there was a window of opportunity – that I'm afraid is closing down now – where people in responsible positions saw that climate change is not going to end in 2050, neither in 2100 nor in 2200, and that we are now still able to give room to the river.

That brings me back to living in the delta. I think we might have an enormous influx of climate refugees from regions – in Europe and beyond – that are facing unlivable conditions. So, sprawling urbanization could be a restrictive element in giving more room to the river in the future. Maybe in 100 years we will not be able to execute projects like this. It will probably be much harder, so there was a window of opportunity for looking at our position in the delta in the very long term. Again, I'm afraid that window is closing slowly now.

Jord: But Dirk, isn't it strange that most people are coming to deltas, although these are the worst places, because the sea level is rising? It's kind of paradoxical. How come?

Dirk: I think it is because the forty or so large deltas of the world – and the Rhine-Scheldt delta is one of them – will be the last stretches of fertile lands. They all were cradles of urbanization in the old days and are still attractive to settle, because they are very flat and have high natural soil fertility. I think they are vulnerable and attractive at the same time.

Jord: In one sense they serve as guarantors for food . . .

Dirk: Yes, and high production . . .

Jord: . . . and we take for granted the price we have to pay for that . . .

Dirk: Yes, and I think that this is a paradoxical situation in deltas across the world.

Audience: Deltas are also very well connected with the global trade system.

Dirk: Yes, I think you're right.

Audience: You mentioned earlier that the Room for the River project was not purely a technical project, but that the people who were in charge valued the spatial quality of the landscape. You said that now that window for projects like this has closed. Is that correct?

Dirk: Yes, I think you are correct. The spatial quality was the second main goal of the entire Room for the River program and that was anchored in the planning of the project. This made the project multi-facetted and integrated in character from the beginning. It also allowed us to bring on board all kinds of designers, landscape architects, and so on. The ambition was to leave the river area more beautiful after the project's completion. After the completion of all the 32 subprojects of the program, policy changed and stripped this approach to water safety from its integral character. From then on, projects had to be strict and sober, following the perception that the quality aspect of the Room for the River program was a 'luxury' – water safety was considered the only goal that mattered. Counterintuitively, this integral and quality-oriented approach made Room for the River one of the very few large-scale projects that stayed within both the planning timeframe and budget. Very, very few large projects achieve that.

Jord: Has it ever been considered that we should retreat, allowing the delta to move inland and stepping back from the first defence line?

Dirk: Yes, and it still is. Retreating is one of the four main scenarios of Deltares, which calculated development and retreatment pathways for our delta. The first time I heard these prognoses outlined was by Henk Saeijs, director and engineer at the service of Rijkswaterstaat, around 1985, at the preparation of a workshop with designers for future scenarios on a national scale for 2050. He challenged designers, saying: 'You need to study the future scenarios in the Netherlands. You're focusing on interventions and investments in the west of the Netherlands, but you should be looking to the high grounds because the west might not be sustainable in the long term.' Well, he foresaw what could happen with the sea level rise. He already said: 'Withdraw.' That will be a discussion for, I think, the next 100 years – because of course there are a lot of people who want to give it a good fight and protect the delta as long as we can.

Jord: But you need money for that. We could afford it in the past, but today there is no income from natural gas, there's no income from colonies like in the 1920s. That kind of funding is no longer available, to recreate the landscape and defend ourselves.

Dirk: This is an existential discussion and a very interesting one. Within the design community, there is a lot of belief that with 'eco-modernist' solutions we will find a way. Futuristic scenarios like enormous dikes around the southwestern delta, that could combine the water safety function with pumped hydro energy reservoirs to store electricity from the offshore wind turbines, and so on. For the consultancy and design companies and the contractors this is an opportunity. But I think these are temporary fixes that will only worsen the problem. The existential discussion about the Dutch delta will be one of the most interesting ones, intertwining politics, engineering, ecology and philosophy.

Research by motion

Sophia: Moving to the second part of our discussion, we wanted to touch upon the topic of research by motion. Through the Reporting the Delta project we aim to use film as a pretext to initiate discussions like the present one, conduct research in a non-traditional way and bring in diverse perspectives and expertise. This approach allows us not only to get acquainted with the Dutch delta heritage but also to test unconventional research tools. How do each of you view the concept of using film as a research tool?

Jord: Since the birth of film at the end of the nineteenth century, our perception of reality transformed completely, particularly our sense of place and time over these 150 years. Film even influenced philosophers like Freud regarding the way in which we perceive and define our being and our existence. Suddenly, we became aware there are so many layers beyond what we see and share. Film has become a very powerful tool to communicate our shared environment. I always mention that almost everyone who visits New York for the first time says: 'I've been here, I know this.' Our notions from films are so strong that reality has become a film. Even now, if we receive films or pictures we think 'I have been here', even though we've not; we look at the small screen and think it's the reality. We are constantly informed and manipulated about reality. Dramatizing it shows a personal opinion on that reality or reveals another story, which makes it even more interesting. We are guided by all these stories to have an opinion on our reality.

Archival material today is growing by the minute: every minute, thousands of films and pictures are made and they're all in the cloud. Who's going to select, out of all that, what reality to trust? In this context, the filmed reality is already considered a lie. But what about the real reality? Can we still find out what the 'real' reality is and communicate it to others?

This is also a challenge for any designer: to have a look at the future, where to start from, with what opinion, with what mentality. Not just looking at the materialization of things, but also shaping future mindsets, which I think is very important.

The latest challenge is artificial intelligence. AI plays a very important role in universities, but also in landscape architecture, in creating the landscape. Many people think that artificial intelligence will take over. I think it's a very interesting, strong tool that collects everything from the cloud and creates something new. It can do a million things better than humans. We are still at the wheel, but I would like to see artificial intelligence working with another artificial intelligence and creating babies for example – real ones, not on the screen.

So, consider all these layers and at the same time that we are human. We are aware of being alive, of our existence. We know we're going to die, but we don't know when or how. We suffer; we have fun. We have to define who we are in this world and that all plays a role when you think about the future of landscape and of architecture. I think that's a very interesting narrative to play with.

Sophia: It seems then that all the film material collected for *Stroom*, with future archival research, could be collected in an instant. How will we then conduct archival research in the future? How do we select?

Jord: Well, there you used the word 'selecting'. Selection has always been essential to life. Deciding which path to take or which information to prioritize has never been straightforward. Selecting and knowing where to find things will be one of the most important things for the future. How can we select? It's not without reason that people become religious nowadays and that they want strong leaders, because they're hopeless. They're standing in this pool of misinformation, of all this information, and they want decisive figures. Someone who points in a certain direction. It's a very political thing and we have to realize that we are constantly archived. We are filmed here [refers to in the room], we know. The moment you enter this building, you know you are registered. Even on the way from the railway station in Delft to here [the faculty of Architecture and the Built Environment at TU Delft] I counted twelve cameras – if I would kill someone now and jump out the window, they could follow me. I'm archived, everything is archived. The flow of the river is archived. It seems that we are in control, but we must remain critical about who holds this control.

Dirk: That's a profound issue. A couple of months ago I attended a lecture in the Eye Filmmuseum in Amsterdam, where two filmmakers took on the challenge of Werner Herzog who said 'not in a thousand years, anybody – not even AI – will make documentaries like Werner Herzog'. And these two filmmakers tried to do so by creating a documentary using AI. Well, it just didn't look the same, they succeeded only partially. They got the voice right, but I'm afraid the film didn't go anywhere yet. Perhaps next year; it may be only a matter of time before AI can do this.

Jord: I can add something to that. Martin Koolhoven, a Dutch filmmaker, was scouting locations in East Asia for a film set in the 1940s and realized AI could generate the setting he wanted through a prompt: '1940s, typical eastern city, very vivid shopping mall, etcetera'. The result was impressive. Today, we could replicate a river landscape in 2030 . . .

Dirk: I want to add something about the value of looking again at old footage. As designers, we have this topo-time travel thing. We can zoom in on topographical maps back at least to 1830 and we can see the ten-year changes. But that's only the spatial element. The important element of Ochse and Van der Wel's movie is the fact that you can see the social side of it, you can almost smell it. You can see the little penniless girls looking for pebbles of coal washed ashore. And it's sort of eye-opening, this social side. Old footage is a unique tool.

Jord: There will be, though, a moment where people will doubt the reality of this footage, thinking that it's not true, it is made by artificial intelligence. There will be doubts about the reality of the past. We have to be aware, because that's a political thing and certain politicians will try to put you in a mood of constantly seeing fraud. There will be a moment when, with all the material available, people will be in constant doubt. This is the danger of it.

Audience: There is a paradox here. Film is becoming dehumanized in a way, although we tried to gain the human experience, something that you can't grasp with other research methods. I was wondering if it still has a future, then? I hear you say now this is going in another direction.

Jord: We can hardly have a view on the future because it is developing that fast. The only thing we have is our five senses and our emotions. So we have to react based on that, and out of this reaction a new one emerges. An instrument like filmmaking can still be a strong tool. Take as an example the low-pace film about the atomic bomb creator, *Oppenheimer*, which is about something that happened 70-80 years ago and is told in a very slow way. It still moves us. You can see in the way the story is told that there is still honesty in it. On the other hand, you have the *Barbie* film, which is complete nonsense, but is fun too. After watching it I wanted to live in a pink house and be Ken.

It all has to do with human emotions and I don't think that will change over time. We still struggle with the fact, as I said, that we are all going to die and we have to deal with that. Within this frame, we have to live and make decisions. So, if you think on that more existential level, I have no doubt that that film can add to that and this is an extra tool to use as research in your design.

Dirk: So, you think that film will have the tendency to go slower and slower? Is it going to be long shots which artificial intelligence won't have the patience to do? Like a Béla Tarr movie, having a 20-minute shot of a man sitting in front of a broken window and a fly buzzing around his head. Will that be the future of film?

Jord: Well, there will be many outcomes. I always talk with my students and ask them: 'Do you read?' And there's always the answer: 'Well, that takes too much time.' And I respond: 'So what? What do you think of life? That takes a lot of time? Reading adds another life to your life, so you live two lives instead of one. You gain time.' Hard to explain. I think fantasy and dreaming are an essential part of our life, and the outcome can be a slow film like, for example, the last Wim Wenders film on Tokyo, which is a fantastic film with constant repetition. But I like any action film or sci-fi film as well. There's so many ways film has developed in the past 100 years that you accept both the fastest and the slowest. I'm not sure what will be next. I am not sure what can be better after film, the fifth art.

Audience: TikTok?

Jord: Well, this is another outcome. A one-minute Béla Tarr. But look at music. Songs – since they were first developed in the 50s – are still at a three-minute span. Three minutes is for us what we consider a song. Well, say that to someone from India. After three minutes, they just get started. After three hours, that's a song. It is a cultural thing too.

Audience: So, perhaps, there is a natural rhythm we seek that goes back to daily experience. What something says to your body, to the rhythm of your heart, the way you feel the environment around you. This is never artificial and it's a good proof of what quality of life is.

Jord: This is a wonderful ending. It brings to my mind a personal story, a memory with my girlfriend. When I first met her, I told her: 'I'll bring you to the place I love'. I grew up in Amersfoort, a very small town, and I brought her there. It looked really bad. She said: 'Why did you bring me here? What do you want to show me?' I said:

'I brought you here because this is the first place where I really fell in love for the first time.' Suddenly this space was filled with memory and the memory is sometimes stronger than the space itself. As a designer, you should recognize this for any design, whether physical or cinematic. The longing for your youth, or nostalgia and genuine emotions. I think that's very important. The beat of the heart.

References

Den Hollander, J. (director, 2008). *De Rijn van Lobith tot aan zee 2007 / The Rhine from Lobith to the sea 2007* [Film]. Jord den Hollander Film, Architectuur Film Festival Rotterdam.

Den Hollander, J. (director, 2005). *Stroom: Collage van het collectieve geheugen / Stream: Collage of the collective memory* [Film]. Jord den Hollander Film

Ochse, I.A. & van der Wel, A.M. (director, 1922). *De Rijn van Lobith tot aan zee / The Rhine from Lobith to the sea [Film].* Polygoon.

Sijmons, D., Feddes, Y., Luiten, E., Feddes, F., & Nolden, M. (2022). *Room-for-the-River: Safe and Attractive Landscape* (Wageningen: Blauwdruk Publishers).

Delta Finale, Joop Burcksen & Ruud Herblot, 1988

Rommy Albers

The film *Delta Finale* is a commissioned film by the Ministry of Transport, Public Works and Water Management. The film was made to mark the closure of the Eastern Scheldt by a pillar dam – the final completion of the Delta Works after thirty years of labour. The title *Delta Finale* refers to Bert Haanstra's film *Delta Phase 1*, the iconic film about the closure of the Veerse Gat in 1961. Not only the title refers to Haanstra's film; the cinematography and content of the film are a direct homage to Haanstra too. Starting with the title sequence, the similarities are unmistakable. What follows is an edit in which the old Zeeland with its traditional costumes and churches is intersected with modern times: high-rise buildings, industrial areas and ships sailing through the canalized landscape. The latter images seem to come straight out of another iconic film: George Sluizer's 1960 film *De lage landen* (*Hold Back the Sea*).

Next, the construction of the pillar dam is shown in detail. The placing of the pillars and the preparatory work, the research in the hydraulic laboratory, the completion of the dam and finally the official opening, which formed the completion of the Delta Works. Besides the Dutch Queen Beatrix and former Queen Juliana and their spouses, high-ranking foreign guests were also invited, including King Boudewijn of Belgium, German President Richard von Weiszacker and his French counterpart François Mitterrand, and members of the British Royal family. The Netherlands wanted to show the world that it had completed its new 'eighth wonder of the world'. No longer was Jacob van Campen's Amsterdam city hall our national wonder of the world; now it is the waterworks that protect the country. And with that comes a cinematic document that, in the full tradition of Dutch documentary filmmaking, captured this final chapter for eternity.

References

Burcksen, J. & Herblot, R. (directors, 1988). *Delta Finale* [film]. Mundofilm.

De Haas, J. & Franken, M. (directors, 1945). *Walcheren* [film]. Nederlandse Werkgemeenschap voor Filmproductie.

Haanstra, B. (director, 1950). *Spiegel van Holland / Mirror of Holland* [film]. Forum Film.

Haanstra, B. (director, 1953). *Dijkbouw / Dike Construction* [film]. Forum Film.

Haanstra, B. (director, 1955). *. . . En de zee was niet meer / . . . And the Sea Was No More* [film]. Bert Haanstra Films.

Haanstra, B. (director, 1955). *Strijd zonder einde / The Rival World* [film]. Bert Haanstra Film/Shell Film Unit.

Haanstra, B. (director, 1958). *Fanfare* [film]. Bert Haanstra Film.

Haanstra, B. (director, 1958). *Glas / Glass* [film]. Sapphire Film.

Haanstra, B. (director, 1962). *Delta Phase 1* [film]. Bert Haanstra Films.

Haanstra, B. (director, 1962). *Zoo* [film]. Bert Haanstra Film.

Mees, J. (director, 1954-1955). *Westenschouwen, zomer 1954 en Zierikzee, zomer 1955* [film].

Mullens, W. (director, 1922?). *Doode steden aan de Zuiderzee/Dead Cities on the Zuiderzee* [film]. Haghe Film.

Roosdorp, A. (director, 1938). *Wieringermeer* [film]. Marofilm.

Roosdorp, A. & Roosdorp-van den Berg, M. (directors, 1945). *Herwinnen door werken / Reclaiming by Works* [film]. Marofilm.

Rutten, G. (director, 1934). *Dood water / Dead Water* [film]. Nederlandse Filmgemeenschap.

Sluizer, G. (director, 1960). *De lage landen / Hold Back the Sea* [film]. Shell Film Unit.

Van Gasteren, L. (director, 1960). *Een nieuw dorp op nieuw land / A New Village on New Land* [film]. Spectrum Film.

Van Gasteren, L. & Meerman, J. (directors, 2012), *Terug naar Nagele / Back to Nagele* [film]. Spectrum Film.

Winar, E. (director, 1938). *De laatste dagen van een eiland / The Last Days of an Island* [film]. Filmbureau Niestadt.

Authors

Joost Adriaanse is an independent design researcher and academic, performing hands-on inquiries into the conditions and possibilities of life in the one world we all inhabit. His fields of interest and expertise involve environmental humanities, cultural analysis, human-non-human relations, circular ecology, and cohabitation. Through action research, writing, filming, prototyping and translating he explores eco-social futures. He works as project initiator, design researcher, educator, writer, reporter, seaweed farmer and free thinker. His key projects include -zee -plaats -werk -land, RE-source (winner of Best Design Research DDA 2019), and Nieuw Zeeuws Landschap Weids en Intiem. In addition to running his design studio, Joost was a tutor at Design Academy Eindhoven, researcher at Vrije Universiteit Amsterdam, and guest teacher at Roosevelt University College and ArtEZ.

Rommy Albers is senior curator at the Eye Filmmuseum in Amsterdam. His fields of expertise are Dutch film history and experimental film. He is editor of the book *Film in Nederland / Film in the Netherlands* (Amsterdam-Gent 2004) and numerous articles on Dutch film history and experimental film.

Sophia Arbara is an architect and urban designer whose work spans from urban design, landscape architecture and cultural heritage studies. Her research addresses socio-ecological challenges through spatial strategies that reveal both known and hidden narratives. As a postdoctoral researcher in Delta Urbanism, TU Delft, she explores food systems as spatial agents in delta regions and the role of soil in spatial planning. She earned her PhD from Roma Tre University and has taught and conducted research at UC Berkeley and the Federal University of Santa Catarina. Sophia previously worked in the Netherlands, Greece, and Spain. She holds master's degrees in architecture from the National Technical University of Athens and in urban design from UC Berkeley.

Nina Bačun is a designer, researcher, and educator exploring the domains of exhibition and set design, conceptual and speculative design and film. Her collaborations with the Zagreb-based collective OAZA reflect her holistic and multidisciplinary approach. She obtained her MA in Design at the Faculty of Architecture, University of Zagreb in 2007 followed by a second MA in Experience Design from Konstfack University of Arts, Crafts and Design in Stockholm in 2011. Nina is currently pursuing doctoral studies in Architecture and Urban Planning in Zagreb, focusing on the intersection of architecture, design, the built environment, and moving images.

Guido Borelli is associate professor of urban sociology at IUAV University of Venice. His research focuses on urban political economics, Marxist critiques of urban policies and the relationship between urbanization and everyday life. He has published books and articles on the urban political economy of Italian, European, and American cities. Additionally, Borelli has authored numerous national and international publications on the work of French philosopher Henri Lefebvre. He curated the Italian translation of Lefebvre's last book: Éléments de rythmanalyse (1992).

Carissa Champlin is assistant professor of transdisciplinarity in the geosciences at the University of Twente. Her research engages design and diverse futuring approaches in the collective planning of resilient cities and urban regions. The inclusion of epistemic diversity in planning processes is central to her work, which integrates evidence and the imagination through city gaming and the prototyping of new geo-information technologies. Her research contributes to various transdisciplinary climate resilience and design research networks in the Netherlands and she serves on the advisory board of Play the City. She holds a Master's degree in urban planning from TU Berlin and a PhD in planning and the built environment from the University of Twente.

João Cortesão is an expert in urban climate resilience driven by the urgency to accelerate the transition to a resilient future. He has experience in developing and implementing climate-responsive design measures and policy advice within the disciplines of architecture, urban design, landscape architecture and spatial planning. João believes resilience is about collaboration and impact, which is why he leverages his experience to tackle the complexity of integral projects that connect climate with other topics. With 17 years working internationally in both academia and the business sector, he pursues this mission by applying scientific knowledge and methods, orchestrating the integration of disciplines, sectors, scales and stakeholders. João has a publication record and is often invited to deliver lectures and talks in different international organizations.

Giuseppe Ferrari and **Nicoletta Traversa** are the founders of RI-PRESE Memory Keepers, a home movies archive dedicated to preserving the audiovisual heritage of the Venetian territory. Since 2013, RI-PRESE has focused on collections by collecting, digitizing, and celebrating Venice's home movies. These materials, spanning formats like 9.5mm, 8mm, 16mm, Super 8, VHS, and Video8, capture life in Venice as far back as the 1920s. RI-PRESE operates on the belief that these personal and intimate stories, often overlooked in official histories, form a rich mosaic of the city's cultural identity. By preserving family memories, the archive contributes to resisting the erosion of Venice's fragile and unique heritage. Beyond preservation, RI-PRESE promotes research, education and creative experimentation. Through the reactivation and sharing of collected footage, the archive fosters new ways of understanding and connecting with Venice's past.

Lia Franken-de Vries is now retired and spent her youth in Bergen op Zoom. After moving to the Kop van Noord-Holland, she studied labor market policy and personnel policy and worked at employment agencies and later as a department head at a smaller municipality. From 1994 to 2014 she was politically active, first as a municipal councilor and the last 12 years as an alderwoman.

Jord den Hollander has a Master's degree in architecture from TU Delft and training in scenario writing from the London Film School. Throughout his career he has often combined both disciplines to make award-winning documentaries about art and architecture. He has written and directed acclaimed television series for children about art and science. In 2001, he co-founded the Architecture Film Festival Rotterdam, the world's biggest architecture film festival. In addition, he has designed a unique oeuvre that includes children's architecture centre Klokhuis, a mobile kid's library, a floating hotel, a television studio from used cattle feed silo's and a tensegrity bicycle bridge. His films have been awarded at many international festivals. He is a much asked lecturer and jury member at international festivals and film and architecture institutions.

Fransje Hooimeijer is an associate professor in Environmental technology and design from the department of Urbanism at TU Delft. Fransje specializes in system integration of technical conditions in urban design and interdisciplinary design. In her research and teaching, she takes the perspective of the city as a technical reconstruction of the natural landscape and investigates how these natural conditions can be utilized better, and how urban systems can adapt to new technologies, through urban design. The focus lies on design and planning processes and how, by representation of the city as a technical construction, it can be made climate-proof, energy-neutral, more biodiverse, and of better spatial quality.

Luca Iuorio graduated in architecture from IUAV University of Venice, where he also earned a doctorate in urbanism. He is currently assistant professor at TU Delft, working on the spatial design of future deltaic systems within the Environmental Technology and Design section and the Delta Urbanism research group. His research focuses on the interplay between engineering, society, and climate, examining how spatial transformations are shaped by human behavior, technology, and environmental forces. He approaches climate adaptation as both a spatial and social process, using qualitative methods to understand lived experiences of change. Luca explores water infrastructures not just as technical solutions but as drivers of urban transformation, bridging urbanism, engineering and environmental sciences to rethink how we design with uncertainty.

Gaby S. Langendijk is a research scientist focusing on developing reliable and robust climate information and services, and supporting decision making processes and planning for climate resilient futures, particularly in urban areas. With a decade of experience, and through close collaborations with cities, she is specialized in climate adaptation, urban climate change and resilient pathways planning. While working at the UN World Climate Research Programme (WCRP), the Climate Service Center Germany (GERICS) and Deltares, she contributed to global, European and national projects. Gaby connects research and practice through interdisciplinary and collaborative approaches, including exploring the interface between science and the arts to drive effective climate action.

Marilena Mela is an architect and assistant professor in heritage studies at Vrije Universiteit Amsterdam. She holds a PhD from the same university, awarded for her thesis titled "Making the Archipelago: Heritage, Energy, Planning and Action in the North Sea and the Mediterranean". Her research investigates the nexus between heritage, sustainability, landscape, and climate – more broadly examining relationships between inherited pasts and imagined futures. She teaches about heritage, climate, landscape, design and the city. She also collaborates with the collective Boulouki, whose work centers on traditional knowledge and building practices in the Greek landscape.

Floris Paalman, PhD, is assistant professor in the Department of Media Studies, University of Amsterdam. He is the coordinator of the MA programme "Preservation & Presentation of the Moving Image", teaches courses on archiving, curating, media history, and research methods. He holds degrees in filmmaking, cultural anthropology, and media studies, and has a special interest in architecture and urbanism. His current research is focused on archiving and curating political cinema and the ontology of audiovisual media archives.

Agnès Patuano is assistant professor in landscape architecture at Wageningen University where she focuses on connecting research and practice through landscape design. She holds an MSc in landscape design & engineering from the Ecole Nationale d'Ingénieurs de l'Horticulture et du Paysage (ENIHP) and a PhD from the University of Edinburgh. Recognized for her work on landscape preference and fractal geometry, her aim is to develop evidence-based methodologies to guide the design of mixed urban spaces that promote healthy lifestyles and behaviours. She has contributed to understanding how landscape design affects physical and mental well-being. Passionate about creating accessible outdoor spaces for all, she strongly believes that well-designed everyday urban environments are essential to fostering healthy, resilient communities.

Jan-Philipp Possmann studied theater- and political- science at Freie Universität Berlin. As a dramaturg, producer and curator he has worked internationally with numerous artists and developed and curated various programs for theaters and museums in Berlin, Frankfurt, Stuttgart, Mannheim and other cities. From 2017 to 2022 he was artistic and managing director of Zeitraumexit and Wunder der Prärie-Festival in Mannheim. Since 2022 he has been developing an agency and service range using cultural heritage to help communities deal with climate adaptation challenges. He lives in Mannheim.

Pınar Şefkatli is an architect and social scientist specializing in rhythms. She holds a PhD from the University of Amsterdam where she explored rhythmanalytical methods for urban research and practice. As a postdoc in urban planning at UvA, she further develops this approach within the context of socially-inclusive climate adaptation. She previously worked for the City of Amsterdam, TU Delft, and architecture bureaus in Amsterdam and Rotterdam.

Saman Seyff is an architect, urban designer and amateur filmmaker. He is currently pursuing a PhD at TU Delft on Tehran's leisurescape, particularly clubs between 1963 and 1979. With Master's degrees in urbanism and strategic planning from KU Leuven, he specializes in architectural humanities, human-centred design, and placemaking. His diverse international portfolio has garnered him recognition and several awards.

Paul Shepheard is a writer and trained architect based in London. Since the publication of *What is Architecture?* (MIT Press, 1994), he has focused primarily on writing and lecturing. His work explores architecture, landscape, technology, and utopia through essays, fiction, and poetry. Shepheard is the author of several books, including *The Cultivated Wilderness* (1997), *Artificial Love*

(2003), *How To Like Everything* (2013), *Buildings: Between Living Time and Rocky Space* (2016), and *Slogans and Battlecries* (2020). His autobiography was published by Canalside Press in 2025. He has taught at institutions including the Architectural Association (London), the University of Texas at Austin, the Academie van Bouwkunst (Amsterdam), and Artesis (Antwerp).

Dirk Sijmons is a Dutch landscape architect known for integrating ecology with spatial planning. He studied at TU Delft and worked for Dutch ministries before leading the Landscape Architecture Department at the Forestry Commission (1984-1990). He co-founded H+N+S Landscape Architects, a leader in sustainable design. He was professor of Landscape Architecture at TU Delft and curated the 2014 International Architecture Biennale Rotterdam (*Urban by Nature*). His publications include *Landscape and Energy* (2014) and *Landscape* (2002). Recognized for his impact, he received the Rotterdam-Maaskant Prize (2002), the Edgar Doncker Prize (2007), and the Sir Geoffrey Jellicoe Award (2017). In 2004, he was appointed the Netherlands' first State Landscape Advisor. His work continues to influence climate adaptation and energy transition in landscape design.

Kamar Ahmad Simon crafts narratives blurring fiction and reality from the Bengal delta. His debut *Are You Listening!* opened DOK Leipzig and won the Grand Prix at Cinéma du Réel. His second feature, *Day After...*, premiered in IDFA's competition, won Open Doors (Locarno), Harrell Award (Camden), and aired on ARTE's *La Lucarne*. He has been featured at Cannes Cinéfondation, Vancouver, Montreal, Zurich, Shanghai, and Sydney; and exhibited at MoMI (New York), Centre Pompidou (Paris), and the RAI (London). Backed by Berlinale, Sundance, IDFA, and Göteborg, his works are archived by Bibliothèque nationale de France and Johns Hopkins. As juror and mentor, Kamar bridges cinema and academia, amplifying marginalized voices – even within the periphery of the 'other'.

Digna Sinke graduated from the Dutch Film Academy in 1972 as a writer and director of fiction and documentary films. Her debut feature, *The Silent Pacific*, was selected for the 1984 Berlinale. In 2010, she completed a long-term documentary on Tiengemeten, a Dutch delta island transformed into 'wild' nature. Her hybrid film *After the Tone* (2014) and the documentary *Keeping & Saving or How to Live* (2018) were both selected for the International Film Festival Rotterdam, with the latter also nominated for Best Feature Documentary. She produces artistic films and has received several honors, including a knighthood in the Order of Orange Nassau (1998), the Groeneveld Prize (2016), and the Golden Calf for Film Culture (2023).

Peter Veer grew up in a professional fruit nursery in the Betuwe, the eastern part of the Dutch Rhine delta, during a transformative era of land consolidation. He studied landscape architecture at Wageningen Agricultural University and spent 25 years directing documentaries on urban and rural development, architecture, landscape, sustainability, and heritage. In 2020, he earned a PhD with "Bewogen Landschap", a cultural-historical study of Dutch Ministry of Agriculture films (1945-1985), a collaboration between the University of Amsterdam and the Netherlands Institute for Sound and Vision. Now an independent documentary maker and rural film historian, Peter lives in central Amsterdam while staying closely connected to the countryside and the quests in the agricultural sector of today.

Ester van de Wiel is an independent designer, researcher, and program maker working in public space and the collective domain. Her work explores socio-material cultures, time coalitions, co-productions, circular ecologies, and cohabitation. Through hands-on methods and empathic fieldwork, she collaborates with human and non-human participants, shifting between close observation and systemic reflection. Her approach combines participant observation, prototyping and dialogue to foster relationships and collective insight. Key projects include *RE-source* (Best Design Research DDA 2019), *-zee -plaats -werk -land*, and *Between Realities* (Best Curatorial Team PQ 2014). She currently tutors at Willem de Kooning Academy and has previously taught at the Sandberg Institute, Design Academy Eindhoven and PolyU Hong Kong.

This publication is part of the project 'Reporting space, time, and everyday life in the delta', funded by the TU Delft Climate Action Programme (Seed Fund 2022) and organized within the Delta Urbanism research group, the section of Environmental technology and design, the department of Urbanism from the faculty of Architecture and the built environment, TU Delft.

The project was curated by Luca Iuorio, Sophia Arbara, and Carissa Champlin, with Feike Smithuis, Amber Coppens, and Merel Garritsen, and developed in collaboration with Rommy Albers of the Eye Filmmuseum, Amsterdam.

This publication was made possible through the financial support of the TU Delft Climate Action Programme and TU Delft OPEN Publishing, and benefited from the collaboration of the Eye Filmmuseum, Amsterdam.

Credits

Editors: Luca Iuorio (TU Delft Urbanism), Sophia Arbara (TU Delft Urbanism), Carissa Champlin (TU Delft Human-centered design & UT Urban and Regional Planning)
Contributions: Joost Adriaanse, Rommy Albers, Nina Bačun, Guido Borelli, João Cortesão, Giuseppe Ferrari, Lia Franken-de Vries, Fransje Hooimeijer, Jord den Hollander, Gaby S. Langendijk, Marilena Mela, Floris Paalman, Agnès Patuano, Jan-Philipp Possmann, Pınar Şefkatli, Saman Seyff, Paul Shepheard, Dirk Sijmons, Kamar Ahmad Simon, Digna Sinke, Nicoletta Traversa, Peter Veer, Ester van de Wiel
Advisory Board: Roy Bendor (TU Delft Human-centered design), Patricia Pisters (UvA Media studies), Marjolein van Esch (TU Delft Urbanism)

Copyediting: Saskia Roselaar
Design: Joseph Plateau
Cover Design: Rens Muis, 75B
DTP: Perfect Service
Visual essay design: Luca Iuorio
Lithography: Pieter Reinink, PRDigitaal

Printing: Veldhuis Media
Production: Sara Duisters, nai010 Publishers
Publisher: Marcel Witvoet, nai010 Publishers

Image credits cover:
...*En de zee was niet meer*, Bert Haanstra, 1955, Bert Haanstra Films. Eye Filmmuseum.
Wakers en dromers, Annette Apon, 1994, Ariel Film. Eye Filmmuseum.
De lage landen, George Sluizer, 1960, Shell Film Unit. Eye Filmmuseum.
Delta Phase 1, Bert Haanstra, 1962, Bert Haanstra Films. Eye Filmmuseum.
Een nieuw dorp op nieuw land, Louis van Gasteren, 1960, Spectrum Film. Eye Filmmuseum.

nai010 publishers is an internationally orientated publisher specialized in developing, producing and distributing books in the fields of architecture, urbanism, art and design. www.nai010.com

nai010 books are available internationally at selected bookstores and from the following distribution partners:

North, Central and South America – Artbook | D.A.P., New York, USA, dap@dapinc.com

Rest of the world – Idea Books, Amsterdam, the Netherlands, idea@ideabooks.nl

For general questions, please contact nai010 publishers directly at sales@nai010.com or visit our website www.nai010.com for further information.

Printed and bound in the Netherlands

ISBN 978-94-6208-913-6
NUR 648
BISAC ARC008000
THEMA TNF

Reporting the Delta is also available as e-book:
ISBN 978-94-6208-920-4 (e-book) and open access through TU Delft OPEN Publishing